PRAISE FOR *POWER LINES*

"The environmental movement and the labor movement need to not just work together. They need to overlap and pollinate each other, with the tools for organizing and analysis that each possesses. This compelling anthology offers a wealth of advice on how to make that task happen—quickly, one hopes, for the sake of a planet that works in every sense of the word."

—Bill McKibben, author of *The Flag, the Cross, and the Station Wagon*

"This is a book that could brighten your life and stiffen your spine. These experienced and wise organizers search the world we share for the stories of movement uprisings that could spark something big enough to save us yet. Read the book. It will make you feel better. It may even make you join the fight."

—Frances Fox Piven

"*Power Lines* presents critical case studies on advancing all communities toward a just transition. The book provides key insights directly from the front lines on how we can organize our communities toward collective power, navigate tensions, and truly advance change. This book makes it more apparent the critical role the labor plays and needs to play in advancing a just transition."

—Shantell Bingham, organizing director of Climate Justice Alliance

"This is it. We are in the midst of one of the most pivotal struggles of the modern era. To meet the challenge of the moment, our movement—both community and labor—must partner and fight together. The essays in *Power Lines* provide powerful examples of how organizers from across the country have taken the challenge of weaving together the strands of our movement for the collective power we need to save us, to save the world."

—Lauren Jacobs, executive director, PowerSwitch Action

"It is widely acknowledged that, if a movement to confront the climate catastrophe is to be successful, it must include working people on the front lines. Often, however, the conversation does not get much further than that. Ordower and Zafir's *Power Lines* shows that we do not need to start at square one in building a labor–climate justice movement. Rather, we can draw rich lessons from organizing that has taken place over the past two decades."

—Mark Engler and Paul Engler, authors of *This Is an Uprising*

"The climate movement needs the labor movement to win a just transition. *Power Lines* is an essential how-to manual for organizers looking for the most creative, visionary, and practical strategies to bridge our movements."

—Varshini Prakash, executive director of Sunrise Movement

"What grabbed me about *Power Lines* was that the analyses offered truly reflect an understanding that climate politics is a struggle over power. This unusually good collection of working class–based environmental justice analyses and summations is thought-provoking and extremely useful in order to advance discussions within social, economic, and environmental justice movements."

—Bill Fletcher Jr., former president of TransAfrica Forum
and co-founder of the Black Radical Congress

"*Power Lines* answers a crucial question: how to make a living on a living planet? With numerous examples to draw from, this book is a necessary corrective to the idea that creating economic prosperity for workers and communities comes at the expense of climate action. In fact, pursuing both goals is not only possible but happening all around us!"

—May Boeve, executive director of 350.org
and 350 Action Fund

"*Power Lines* is a powerful book that highlights how increasing numbers of unions are engaging in deep and important political education work with their members about the role of their employers in profiting by ravaging the planet. The next step is to make climate issues a central bargaining issue by Bargaining for the Common Good, creating concrete demands that protect workers, their communities and the planet."

—Stephen Lerner, labor and community organizer
and architect of the Justice for Janitors campaign

"Decades of collective experience inform the writing in *Power Lines*; key organizers connect the dots of struggle in the sometimes fraught intersection of labor organizing and movements for climate justice. We can learn lessons from both success and failure, and this collection clarifies the central question that must animate our organizing: how do we build the kind of power that can help ordinary people re-shape the world?"

—Alex Han, *In These Times*

Jeff Ordower is an organizer with over thirty years of community and labor organizing experience. He is currently the North America director of 350.org. Prior to joining 350, he was a co-founder of the Green Workers Alliance. Ordower learned his base-building organizing skills through sixteen years in various roles at ACORN.

Lindsay Zafir is a distinguished lecturer at City College of New York and the academic director of Leadership for Democracy and Social Justice. She is the former editor of *The Forge: Organizing Strategy and Practice.*

POWER LINES

Building a Labor-Climate Justice Movement

EDITED BY JEFF ORDOWER
AND LINDSAY ZAFIR

NEW YORK
LONDON

Published in the United States by The New Press, New York, 2024
Distributed by Two Rivers Distribution

ISBN 978-1-62097-818-4 (pb)
ISBN 978-1-62097-822-1 (ebook)
CIP data is available

The New Press publishes books that promote and enrich public discussion and understanding of the issues vital to our democracy and to a more equitable world. These books are made possible by the enthusiasm of our readers; the support of a committed group of donors, large and small; the collaboration of our many partners in the independent media and the not-for-profit sector; booksellers, who often hand-sell New Press books; librarians; and above all by our authors.

www.thenewpress.com

The Forge

Power Lines was developed as a project of *The Forge: Organizing Strategy and Practice*

Composition by dix!
This book was set in Garamond Premier Pro

Printed in the United States of America

10 9 8 7 6 5 4 3 2 1

CONTENTS

Preface ix
Matthew T. Huber

Introduction: Our Future Story 1
Miya Yoshitani and Jeff Ordower

Part I. Our Work Makes the System Go: Organizing Labor for a Just Transition

1. The Dream and the Nightmare: Organizing Oil Workers for a Renewable Energy Future 19
Norman Rogers

2. "Our Work Is What Makes the Food System Go": An Interview with Edgar Franks 27
Jeff Ordower

3. Pushing for a Green New Deal for Education from Below 35
Todd E. Vachon

4. Care Work Is Central to a Just Transition 53
Batul Hassan

5. Young Workers Can Bridge the Labor and Climate Movements 61
Maria Brescia-Weiler and Liz Ratzloff

6. Organizing Climate Jobs Rhode Island 73
Patrick Crowley

Part II. Killing the Wiindigo: Climate Justice, Worker Justice

7. How to Win a Just Transition: An Interview with Jose Bravo 83
Jeff Ordower

8. Organizing Coal Country 91
Veronica Coptis

9. "Resilient Communities Are Organized Communities": An Interview with Vivian Yi Huang and Amee Raval 101
Miya Yoshitani

10. Good Jobs, Clean Air: How Community, Environmental Justice, and Labor Groups Stopped Amazon's Air Hub in New Jersey 111
Sara Cullinane and Wynnie-Fred Hinds

11. Killing the Wiindigo: Restoring Interdependence and Uniting Our Movements 123
Winona LaDuke and Ashley Fairbanks

Part III. "We Are Our Best Chance for Rescue": Organizing the Front Lines of the Climate Crisis

12. "We Are Our Best Chance for Rescue": Green Workers Organizing for More and Better Jobs 135
Matthew Mayers

13. Solidarity for the Snowpack 145
Isabel Aries, Ryan Dineen, and Katie Romich

14. The Fight for America's Workers Must Also Be a Fight for Environmental Justice 153
Tefere Gebre

15. Don't Waste LA: An Interview with Lauren Ahkiam 165
Miya Yoshitani

16. Building a Worker-Led Movement for Climate Justice 171
Brooke Anderson

17. Listening to the Land, Listening to the Workers: Farmworker Organizing on the Front Lines of Climate Crisis 187
Davida Sotelo Escobedo, Max Bell Alper, Davin Cárdenas, and Aura Aguilar

Acknowledgments 201

Notes 203

PREFACE

Matthew T. Huber

The cause of labour is the hope of the world.[1]
—Walter Crane, *The Workers' Maypole* cartoon

In a time of global climate breakdown, it is worth recalling the worldly outlook of the traditional socialist and labor movements. Much of the workers' movement saw species-wide emancipation as the ultimate goal of struggle ("the internationale unites the human race"). Now, of course, climate change calls for a similar planetary movement with species survival at its core. But, as it turns out, emancipation from capital is just as necessary to solve this crisis as workers and socialists believed it was a century ago.

Too often climate change is presented as a struggle over science and knowledge or as a technocratic process of implementing policies that channel market incentives. As this volume illustrates well, climate politics is a struggle over *power*. There are immense vested interests in the capitalist class whose power and projected profits over our energy, agricultural, transportation, and other systems rest on maintaining carbon-intensive forms of production for decades to come. If society is to embark on climate solutions at the speed and scale required, we also need to develop forms of social power capable of confronting those interests.

This is why labor and the working class must be central to the climate struggle. Under capitalism, no other social group possesses the same kind of *potential* power as the working class. It is not only their power in numbers as the vast majority of society, it is their strategic power at the point of production. The fact is that it is workers who keep society running and capital's profits flowing. It is this basic fact that gives workers unmatched strategic leverage. When workers go on strike, it creates a crisis that power *must* respond to. As we saw in 2022 in the case of the rail workers in the United States, even the mere *threat* of a strike sent President Biden and the rest of the American power structure to work overtime to prevent it (sadly, it was Congress, including many progressives, that used its legislative power to break the strike).[2] In 2018, in West Virginia, when teachers went on strike and shut down a core institution of social reproduction (schools), the crisis forced a right-wing political system to respond and grant the teachers' demands in a matter of weeks (imagine if they had attempted to win these demands through lobbying or getting the right people elected).[3]

The climate crisis requires a working-class politics, but this volume shows well that the working class is not homogenous and that their "interests" in climate action are highly differentiated. It's possible to offer a rough (and certainly not exhaustive) typology of these differently situated workers. First, there are *low-carbon* workers, like teachers, nurses, and others in the "care" sector, who tend to align with climate action in general and with the expansion of low-carbon care sectors in particular.[4] After all, teachers' and nurses' unions were some of the first to endorse a Green New Deal.[5] Second, there are *climate-vulnerable* workers, like farmworkers and ski workers, whose jobs and livelihoods are threatened by climate change itself. For these workers, climate action is existential (making climate organizing more obviously connected to real material conditions). Third, there are *strategic industrial and energy* workers at the heart of the industrial system that we need to dramatically transform (some call the energy transition a *green industrial revolution*).[6] While climate activists tend to think of these narrowly as "green" workers in the renewable energy industries, I would argue we need to think much more broadly about industrial trade union workers to include much of the building trades and construction industries.[7] Electricians, line workers, carpenters, pipe fitters, painters, and many more—these are the workers we need to *build* the energy transition.

We also must include in this category fossil fuel workers. As this volume points out, Tony Mazzocchi's call for a just transition would actually have to *deliver* real material supports (like free education and at least five years of 100 percent income supports) for these workers to believe a real just transition exists for them.[8] But, the climate left should also consider how these workers possess immense skills and knowledge that could be harnessed toward the green transition. For example, unionized power plant workers could transition from coal and gas to a very similar but low-carbon and reliable nuclear power (a simple reason why electrical unions center nuclear power in their climate recommendations).[9] Oil and gas workers could actually do what Holly Buck calls "reverse engineering": instead of extracting oil they could capture carbon from the atmosphere and use their skill set to inject it back underground.[10]

In closing, I would humbly suggest the climate justice movement is making great inroads with the first *two* kinds of workers—low-carbon and climate-vulnerable workers—but it is hard to imagine large-scale decarbonization at the speed and scale required without the third kind of worker, the strategic industrial worker, on board. Of course, these workers and unions have been notorious antagonists of the climate agenda, but I would suggest that is mainly because the bulk of climate activism for the last several years has focused on "blocking" harmful climate infrastructure (what Naomi Klein famously dubbed "Blockadia").[11] This blocking is important, but it is also true that actually *mitigating* climate breakdown will require a whole lot of *building*: building new clean energy, transmission, public transit, public housing, and agricultural systems. It is very clear this climate politics of building could easily appeal to the material interests of the strategic energy workers at the core of the transition. With all three sets of workers on board, a mass working-class-led Green New Deal could once again seem possible. If that happens, this volume will be a vital resource for organizers.

POWER LINES

INTRODUCTION

OUR FUTURE STORY

Miya Yoshitani and Jeff Ordower

Most of us believe that no matter where we live, workers should be paid a living wage, children should grow up breathing clean air, and all communities should have thriving local economies and healthy environments. But decades of relentless messaging and policy driven by corporate polluters have created the false paradigm of "jobs versus the environment." This paradigm is premised on the assumption that the economic interests of working families are in competition with our shared right to a clean and healthy environment and dignified work.

To take one example, in 2006, after the state of California passed the Global Warming Solutions Act—a historic climate law to reduce carbon emissions—oil companies launched a massive effort to repeal the law based entirely on a "jobs versus the environment" message. Oil companies like Valero Energy and Tesoro spent over $10 million on a ballot initiative—Proposition 23—to delay the implementation of the climate law "until unemployment dropped below 5.5% in the state for a year."[1] In one of the many ads that blanketed the state before the election, a woman says earnestly, "I want to do my part on global warming. All Yes on 23 says is let's wait until people are back to work and we can afford it."[2] California voters rejected Proposition 23, but the oil companies have successfully deployed the same message time and again to block and delay essential environmental regulations at the state and federal levels.

Some unions have played along. The Laborers' Union, for example, has consistently criticized climate activists attempting to halt the extraction and transportation of fossil fuels. At the height of the fight at Standing Rock, when activists shut down five pipelines, the Laborers' District Council of Minnesota and North Dakota released a statement in which they blasted "self-styled environmental activists" for "dangerous acts of sabotage." The Laborers argued that shutting down pipelines would not affect fossil fuel consumption. "Instead, [these actions] only serve to alienate our members and millions of other working people who know exactly how important reliable fuel supplies and other petroleum inputs are to their lives and their livelihood from the cause of fighting climate change."[3]

In recent years, the intensifying climate crisis—coupled with seismic shifts in the energy economy and steadfast organizing by environmental justice and frontline workers' organizations—has shifted the momentum. It now feels possible to advance an agenda for just and equitable climate solutions that create good union jobs, build healthier and more resilient neighborhoods, and generate community wealth in places that need it the most.

For organizers, however, this moment presents an opportunity to do something even bigger: to cultivate the people power needed to win generational fights for racial, economic, and environmental justice—that is, to fight for solutions that match the scale of the problem. That is the only way we will win. Nothing but systemic change in the economy and our democracy will save us from the worst of the climate crisis, and nothing but millions of people engaged in collective action has the power to make that system-level change happen.

The story we hope to tell the next generation is that we survived the climate crisis by building an intersectional movement, across frontline communities and workers, that was powerful enough to transform our society in service of our shared dreams for health, wealth, and self-determination. A story of how the people most impacted by the crises used their knowledge and skills to lead the way. A story of how we won what we wanted and needed when we fought for and with one another on all fronts.

This story will have countless starting points. There have been thousands of sparks across the country over the last few decades alone. Untold numbers of communities of color, tribal communities, and poor rural communities have been sacrifice zones for industrial pollution, targeted by racist policies like redlining and denied basic infrastructure for clean water, healthy food, housing, energy, and transportation. Nonetheless, these communities have stood up, gathered their forces, and started to fight for something completely new. Low-wage workers, often part of these same communities, have been refused the dignity of a living wage, exposed to unsafe or unhealthy working conditions, and denied sick days and health care for their families. Yet they have also been taking collective action to fight for better lives.

We believe that we will look back at these fights not just as inspiration but as instruction. Each one of these skirmishes holds important clues to how we build power. Each one can be studied not just for the win or the loss, the changed policy, or the solution to be scaled but for the exact way each point of contestation helped to engage more people than before and invested them in one another's well-being as essential to their own. In turn, these people built stronger, more powerful, more strategic, and more effective organizations and alliances. Together, they formed movements capable of bringing lasting, transformational change.

Over the long term, we hope these sparks will catch fire into a labor–climate justice movement capable of winning a just transition—a transformation of the extractive fossil fuel economy to a healthy, regenerative, equitable, and democratic economy. The concept for a just transition was first developed by labor leader Tony Mazzocchi, the former vice president of the Oil, Chemical and Atomic Workers International Union (OCAW) (later absorbed into the United Steelworkers). Mazzocchi conceived of a superfund for workers that would "make provisions for the workers who lose their jobs in the wake of the country's drastically needed environmental cleanup."[4] Today, the fight for a just transition is being led by and for workers, communities of color, and low-income communities. These are the people hit first and worst by the triple threat of poverty, racism, and pollution—and they are the creative force behind a vision for safe, healthy, affordable, thriving communities.

Bringing together the grassroots power of frontline communities and frontline workers is not the only way we win on climate. But we cannot win without it. That is why these spark stories, unfinished as they are, are so valuable as we build the movement we need in the time we have left.

Spark Story: Company Town to Transition Town

This is a book of spark stories—a collection from the front lines of the climate crisis about the power we need to win a just transition for workers and communities. One of those sparks is the organizing story of the refinery town of Richmond, California, told through the lens of one of the many community-based organizations that started there: the Asian Pacific Environmental Network, or APEN, where Miya worked for twenty years as an organizer and, later, the executive director.

APEN's story begins with the Laotian refugee community that settled in Richmond during the mid- to late 1980s and helped APEN become a powerful force in putting Richmond on the path to a just transition. Richmond is a racially diverse, working class city that sits eighteen miles northeast of San Francisco—part of the refinery corridor located in the "progressive bubble" of the Bay Area. The Chevron Richmond Refinery was built by the Standard Oil Company a few years before the city of Richmond was established in 1905, and Richmond was a company town from the very beginning. Richmond's population increased dramatically during World War II as Black migration from the South brought thousands of new residents to build ships in the Kaiser Richmond shipyards. As the city's population expanded again in the postwar years with the arrival of Latinx, Asian American, and Pacific Islander immigrant and refugee communities, so too did the industrial polluters. By the 1990s, Richmond's residents were surrounded by over three hundred toxic sites, including the largest contributor to climate pollution in the state, the Chevron refinery. The county that Richmond is a part of is home to the highest number of polluting industrial facilities per capita of any state in the country.[5]

APEN was founded in 1993 to organize working-class Asian American immigrant and refugee communities in the Bay Area to fight for environmental justice. The organization built on the work of partner groups like West County Toxics Coalition and Communities for a Better Environment (CBE). What started as a small organizing project in Richmond has grown over the last three decades into a powerful statewide organization with racial and environmental justice at its core.

APEN's early organizing in Richmond's Laotian refugee community was informed by Mazzocchi's vision for a just transition. In 1996, APEN was part of a project of the Just Transition Alliance (JTA) to bring refinery workers together with "fenceline" community members—people living next to polluting facilities—to surface some of the inherent tension between the two groups, explore common ground, and talk about what a just transition for workers and impacted communities could look like. As part of the process, community members and workers met face-to-face in facilitated conversations that allowed them to listen deeply to each other's perspectives and to humanize their positions. This didn't mean that workers suddenly stopped defending their jobs or community members stopped defending their health, but it allowed them to see where they had deeper shared interests—to look beyond the conflict of the day and imagine a future where they both got what was best for their families and communities.

Three years later, in 1999, four refinery workers were killed during a risky maintenance operation at what was then known as the Tosco refinery, in nearby Martinez. Only a few months after that, the Chevron refinery had a massive explosion, leading to a fire that spewed toxic smoke over much of the northern Bay Area and forced thousands of local residents to shelter in place. The county sent safety alerts and instructions to stay indoors in English, meaning that many of APEN's members were exposed to further danger while they sought information about what was happening. These events inspired APEN's earliest organizing campaign, a fight to get the country's first multilingual emergency warning system to protect the health of immigrant and refugee communities.

APEN spent the next decade organizing around the slower moving but still deadly disasters of racism, poverty, and daily pollution burden. The organization was part of a local coalition of groups that successfully fought an initial $1 billion refinery expansion project to allow the Chevron refinery to process heavier, dirtier crude from places like the Bakken oil field of North Dakota. The campaign brought together local environmental justice organizations and United Steelworkers (USW) Local 5 to promote worker and community safety, an area where the interests of workers and community members often overlap. Environmental justice (EJ) organizations maintained that the Richmond City

Council had rubber stamped Chevron's permitting requests for the refinery expansion without properly considering the environmental, health, and safety impacts, as is required by law. But the tenuous—if not transactional—partnership broke down in the end when the USW International removed its support for the EJ position.

Local EJ organizations like APEN had a more adversarial relationship with the Contra Costa Building and Construction Trades, whose members stood to gain over one thousand construction jobs with Chevron's expansion. This was a classic jobs-versus-environment wedge, with one of the most profitable global corporations on the planet pulling all the strings. Richmond City Council and Planning Commission public meetings, already notorious at the time for being chaotic and openly hostile, became extremely tense and absurdly long. Both sides would line up for hours of public comment inside the meetings, everyone sitting with their chosen side's branded T-shirts. APEN showed up with hundreds of bánh mì, the Vietnamese sandwiches that have always been the organization's direct action food of choice.

There is nothing unusual about confrontations at public meetings such as these, especially when people are divided by strongly held beliefs and when profits are on the line. But looking back, one experience in particular foreshadowed the weaponization of divided interests that has become so normalized over the last decade. APEN's members, mostly seniors and non-English speakers, were confronted by union members twice their size and half their age, who blocked them from entering the building while demanding to know their citizenship status. It was a low point, and one that clearly emerged out of the company's work to pit workers and community members against each other. APEN later found out that Chevron was instructing the union to turn out its members to accuse APEN of taking away their jobs and was paying residents to support its expansion plans in public hearings.

In August 2012, the Chevron refinery exploded after a highly corroded section of old pipe in the crude unit failed and ignited a massive fire. Once again, the refinery had endangered the lives of workers and community members. Nineteen workers narrowly escaped death, and the towering column of black smoke that formed a contaminated dark cloud visible all over the Bay Area sent over fifteen thousand residents to area hospitals with respiratory problems, nausea, and other ailments.

The fire was devastating. But it also catalyzed a decade of organizing and the launch of multiple power-building campaigns, transformative policy initiatives, and new alliances that have had a huge impact on the fight for climate justice in Richmond and throughout California.

One of the first formations to come directly out of the whole calamity was the Refinery Action Collaborative (RAC)—a coalition effort to bring together Bay Area labor and environmental justice groups to protect the health and safety of refinery workers and community members. When the state agency CalEPA established an interagency task force on refinery safety, RAC worked with the West County Toxics Coalition, the BlueGreen Alliance, the Natural Resources Defense Council, Communities for a Better Environment, APEN, and the Labor Occupational Health Program at the University of California, Berkeley to put forward recommendations that held the mutual interests of workers and the community. It was an obvious yet highly unusual move for everyone involved, and the collaboration helped to set the stage for further refinery reforms at the state level.

Ten years later, in 2022, the California Air Resources Board (CARB) approved an updated target of cutting greenhouse gasses by 48 percent below 1990 levels by 2030, with an ultimate goal of cutting oil use by 94 percent and becoming carbon neutral by 2045. The rapid pace of change has shifted the conversation about the future of the Chevron refinery dramatically. What happens next will likely be challenging for workers and the community, but the possibility of building collective power between refinery workers and Richmond residents for a just transition only exists because of the relationships built and tested over the last two decades.

The story of Richmond's just transition is still being written. But the changes that have been made through community organizing—new models for local clean renewable energy, more equitable food systems, affordable housing laws and tenant protections, a progressive tax on big business, stronger local democracy and an expanding progressive voter base, and new state funding for climate resilience and infrastructure—have helped Richmond's social and environmental justice organizations engage more people and expand their base. It is a positive feedback loop that continues to accrue more power to frontline communities and more power for solutions to the climate crisis.

Lessons

The spark stories collected in this volume hold many lessons for the fight ahead. But these lessons and case studies are not useful in a vacuum. We hope that you, as readers, are inspired by these stories and incorporate their lessons into your own organizing.

1. **Workers and frontline communities have the skills and knowledge to lead the way out of climate change.** The people who are building alternatives, creating new infrastructure, planting seeds, harvesting crops, and caring for communities are best positioned to envision a new and regenerative economy—and to bring it into being. "Amidst growing climate chaos," Davida Sotelo Escobedo, Max Bell Alper, Davin Cárdenas, and Aura Aguilar of North Bay Jobs with Justice (NBJwJ) write

in this volume, "Our future will be determined by the leadership of immigrant and Indigenous workers who will do the critical work the Land desperately needs. This means not just fighting to improve current jobs but beginning to actualize a vision of worker-led stewardship rooted in real relationship with the Land." The most effective and resilient solutions to climate change are being led by workers and communities hit the hardest by the problem, and we all benefit when they win.

2. **Frontline communities and workers have the most strategic potential to bring about the fastest shift.** The workers who operate climate-polluting industries—and the communities most affected by these industries—are in the best position to force rapid change. It is essential to build new infrastructure for a fossil-free economy, but it is also imperative to force a transition out of the old and to do so in the most equitable way possible for workers and communities. Organizing these workers and community members to help shape a just transition from fossil fuels is not just morally right but strategically necessary. If we are to build public and political support to force the managed decline of climate-polluting industries, we will need the leadership of the workers and communities at the center of those industries.

3. **Successful labor-climate coalitions are grounded in real relationships, deep listening, and a willingness to lean into conflict.** This lesson emerges in almost every article in the volume. As Veronica Coptis, formerly at the Center for Coalfield Justice (CCJ), writes, "Building a big tent starts with listening to workers, even when

they don't agree with us." CCJ's success at bringing coal workers into climate organizing is due in large part to the organization's commitment to deep listening and relationship building as well as to organizers' willingness to engage in conflict. When a coal mine put out false information that CCJ was trying to shut down the mine—and take away jobs from miners—the organization put up a Facebook post encouraging angry workers to give them a call. "In that moment," Coptis writes, "the workers learned that CCJ operated in good faith and the boss did not." Similarly, many of the members of Green Workers Alliance (GWA) are Trump supporters, but they are fighting for green jobs that can support them and their families. GWA holds space for internal debate around issues like gender, race, and vaccination status, building a big tent while challenging workers on their assumptions. Organizing for climate justice at scale requires that we step into generative conflict that changes our organizational plans as well as the views of members.

4. **The rules of worker organizing—written and unwritten—must be changed.** Too often, these rules get in the way of the kinds of productive experiments that will allow us to build power. One of these rules is that workers only have a union once they've won an election and bargained a contract—a yearslong process made all the more challenging by employer-friendly labor laws. As a result, much of the most exciting and promising organizing to build a labor–climate justice movement is happening outside the confines of traditional unions. The GWA, for example, is directly organizing workers and demanding better wages and working conditions for the

sector. Familias Unidas por la Justicia is experimenting with farmworker-led cooperatives and represents workers not covered by a contract. Workers are in motion all over the United States—building formations that are new and different, outside the structure of traditional labor. We hope that the future will see an increasingly dynamic and freewheeling workers' movement.

The internal structures of unions can also hamper their ability to take on risky organizing projects. Union leaders are elected by their membership. This democratic structure is a good thing, but sometimes it also means that union leaders won't take a stand on issues they think will upset their members. Too often, these decisions are based on assumptions about where the members stand rather than being rooted in organizing conversations. Union members don't live on a faraway planet; they are part of communities affected by toxic pollution and other climate issues.

Other times, unions are cautious about organizing too many people too fast because an expanded membership creates uncertainty. Some unions get stuck on defending the jobs of the past rather than fighting for the jobs of the future—and making them more accessible to more people. In both cases, union leadership must lead. They should do what John L. Lewis, founder of the CIO, did: mortgage the entire union treasury on organizing and lead members into a clear vision of the future, not the past.

5. **Labor is not a monolith.** Community groups can also be overly cautious about approaching labor. An unnuanced view of labor, too often perpetuated by unions

themselves, can leave community organizations with the mistaken belief that enmity or principled disagreement with one union will lead to disagreement with the entire House of Labor. But labor unions represent workers from different sectors with different interests. Community organizations shouldn't assume that taking on one union will mean that they can't align with a different union. As Brooke Anderson reflects on her work organizing union members and organizers around climate change, "You don't have to organize everything and everyone. You just have to move the landscape enough to show cracks in labor's unanimity."

6. **The environmental movement also is not a monolith.** The environmental justice movement should not be conflated with conservation and top-down climate advocacy work. The "Big Green" environmental approach has been dominated by technological and policy fixes that try to isolate the solutions from the root causes of climate change. Advocates of this approach do not have an accurate assessment of how to build power for the solutions necessary to address the scale of the problem. By contrast, environmental justice has its roots in the Civil Rights Movement and is grounded in organizing working-class Black, Indigenous, Latinx, Asian and Pacific Islander, immigrant, and refugee communities. Workers and environmental justice communities are often one and the same, and, despite the fact that we are often pitted against each other, our long-term interests are aligned. We can achieve more powerful and transformational change when labor and environmental justice interests build power together.

7. **The transition will be painful for everyone.** Too often, environmental groups try to downplay the pain that workers and communities will feel as a result of the transition to a renewable energy economy. But we are already feeling the pain of transition. Workers are losing jobs, communities are losing tax revenues, and pollution is getting more concentrated in overburdened neighborhoods. Rather than downplaying the challenge, we should confront the reality head-on and address it by organizing workers and community members around the solutions that really work for them. There would be a lot more solidarity between workers and community members if neither group felt like they were being lied to or sold a story that just isn't true. The transition is inevitable, and it will be difficult. Instead of having the transition happen *to* us, we can be part of organizing around what we need to thrive.

All Roads Lead to Organizing

Environmentalists like to say that climate change is an existential crisis. But the existential crisis of climate change is not about policy; it is not about parts per million of greenhouse gasses, nor is it about innovations in technology. The existential crisis of climate change is about power. Winning the depth and scale of change required to reverse the climate crisis requires real power, and the only source of power available to us is in the number of people we can organize to take collective action—where we live, where we work, where we go to school, where we pray, and where we play.

For decades, the environmental movement has tried to find a shortcut to solving the climate crisis. The perfect market solution, the perfect technology, the perfect policy intervention, the perfect message. Yes, we need these things, but the infrastructure we have never invested properly in is the infrastructure of our movements, the resilience of our leaders, and the ecosystems of our organizing. We can scale people power faster and more ambitiously, but we can't skip over it. There are no shortcuts to organizing for building power.

The stories in this anthology are evidence that communities and workers at the front lines of the climate crisis are starting to find ways to win and build power together. These sparks illuminate a path forward, but we need to choose to do the hard work of organizing every day: knocking on doors, talking to people about their concerns, showing up for one another in times of crisis, standing shoulder to shoulder in the streets, elevating the voices that haven't been heard, and building organizations and alliances that hold and exercise our power for the long haul. We all deserve the dignity of safe, healthy, and well-paid work, and we all deserve to breathe clean air, drink clean water, and live in thriving communities. As much as frontline workers and communities are facing the brink of disaster, we also might be on the brink of a new and better future if we can learn how to organize for it together.

Part I

OUR WORK MAKES THE SYSTEM GO

ORGANIZING LABOR FOR A JUST TRANSITION

1

THE DREAM AND THE NIGHTMARE

ORGANIZING OIL WORKERS FOR A RENEWABLE ENERGY FUTURE

Norman Rogers

I have a dream. I have a nightmare.

The dream is that working people find careers with good pay, good benefits, and a platform for addressing grievances with their employers. In other words, I dream that everyone gets what I got over twenty-plus years as a unionized worker in the oil industry.

The nightmare is that people who had jobs with good pay and power in the workplace watch those gains erode as the oil industry follows the lead of steel, auto, and coal mining to close plants and lay off workers. It is a nightmare rooted in witnessing the cruelties suffered by our siblings in these industries—all of whom had good-paying jobs with benefits and the apparatus to process grievances when their jobs went away. Workers, their families, and their communities were destroyed when the manufacturing plants and coal mines shut down, with effects that linger to this day. Without worker input, I fear that communities dependent on the fossil fuel industry face a similar fate.

This nightmare is becoming a reality as refineries in Wyoming,

Texas, Louisiana, California, and New Mexico have closed or have announced pending closures. Some facilities are doing the environmentally conscious thing and moving to renewable fuels. Laudable as that transition is, a much smaller workforce is needed for these processes. For many oil workers, the choice is to keep working, emissions be damned, or to save the planet and starve.

United Steelworkers (USW) Local 675—a four-thousand-member local in Southern California, of which I am the second vice president—is helping to chart a different course, one in which our rank-and-file membership embraces a just transition and in which we take the urgent steps needed to protect both workers and the planet. Along with other California USW locals, we are fighting to ensure that the dream—not the nightmare—is the future for fossil fuel workers as we transition to renewable energy.

The story of Local 675 is very much tied to its history as a local under the former Oil, Chemical and Atomic Workers Union (OCAW). The term *just transition* was coined by OCAW member and leader Tony Mazzocchi, who saw the harmful conditions that his members labored under in atomic plants. (One of those members was the chemical technician and whistleblower Karen Silkwood, who died in a car crash in 1974 on her way to deliver documents to the *New York Times*.) Mazzocchi envisioned a path forward—a just transition—where workers could find safer employment at the same high wages and good benefits. Mazzocchi's vision for a just transition emerged out of his broader political work. He helped found Labor Party Advocates (with the goal of forming a Labor Party in this country), and he was a strong supporter of the New Majority Party, the precursor to today's Working Families Party.

We are doing our best to follow in Mazzocchi's footsteps by preparing our local for the needed transition to a regenerative economy. If you believe the fossil fuel companies, oil production will remain stable for years to come. That may or may not be the case, so our job—our responsibility—is to prepare our members for what might be next and to fight to ensure that their interests are represented in the transition.

The first step is to get our members and leaders to understand the future of the oil sector. The oil companies hold town halls and informational meetings with workers in which they trot out charts and graphs forecasting that oil demand will remain strong. But when the workers press for specifics, those in charge don't have a good answer. For instance, in one town hall, a representative for a multinational oil company working in Southern California put up a chart showing the projected global demand for fossil fuels moving forward. When one of our members asked about what demand looked like for California, they had no answer. We developed our own clear-eyed presentation, one that emphasizes the need for a transition and gives our leaders and members an opportunity to discuss the trends. But just the fact that the bosses are holding these meetings has been an education for workers. If the company cares enough about the future of oil to educate the workers, then they are obviously worried.

Second, we need to fight for government funding to support fossil fuel workers who lose their jobs during the transition. Los Angeles city and county governments have developed a just transition task force for extraction workers set to lose their jobs as oil wells close. Both the city and the county passed legislation mandating that oil and gas wells be a certain distance from homes and schools, leading to the closure of a number of wells and subsequent job loss for workers. In December of 2022, that task force released a blueprint for how workers and communities can transition toward a regenerative economy.[1]

Across the state, a dozen unions have joined together in

California Labor for Climate Jobs to push for a just transition for fossil fuels. Together, we successfully lobbied the state of California for a $40 million fund for displaced oil and gas workers and a $20 million fund for displaced extraction workers. Marathon Oil, for example, has already begun converting to renewable diesel, resulting in significant job losses for members of USW Local 5 in Northern California. Job loss is especially challenging for older members who are too close to retirement for retraining. The funds are designed to support workers through retraining programs, early retirement, wage replacement, mental health services, and certification of work experience and training. Implementation is always key, so we are working hard to ensure that local and state governments are getting this money out the door and into the hands of workers as soon as possible.

The funds resulted from a joint effort between labor and environmental organizations, which realized good jobs must be part of any transition to renewable energy. These kinds of coalitions are not always easy to build or sustain. Working together required labor to educate environmental groups on the range of essential goods that come out of fossil fuel production—from vaccines to clothing to household appliances—as well as to organize our own members on the inevitability of change and the need to plan for it. The transition is happening, and we have to be nimble in our response.

Among the obstacles we are facing is a history of bad U.S. industrial policy, which makes workers skeptical that any transition will actually be a just one. One only need visit Youngstown, Ohio, or Detroit, Michigan, to see the lasting effects of the demise of steel and auto, respectively. Workers in the oil industry do not want to see their towns abandoned and their livelihoods gone with a transition to renewable energy.

Fossil fuel companies pose another hurdle. The fossil fuel

industry has money and access. At the Climate Change Conference (COP 27) negotiations in Sharm el-Sheikh, Egypt, in November 2022, the fossil fuel presence was overwhelming, with over six hundred lobbyists in attendance. The talks resulted in funding for loss and damages among countries that have not contributed to climate change but suffer disproportionately from its effects. This is a worthy goal but one that maintains the status quo in regard to future emissions.

Obstacle three: union leadership—as well as our rank-and-file members—are unsure about how to maintain the gains we have won over so many years. Many members feel affinity to the companies that have provided us with work, and rightfully so. The fossil fuel industry has provided steady employment and, through collective bargaining, paid good wages and offered decent benefits. Fossil fuel jobs have launched generations of workers into the middle class and sometimes beyond. If our employers say all is well, it is difficult to fight them because our members want to keep their jobs. That is exactly what the companies will do right up until the time we are no longer needed.

The fossil fuel industry has long seemed impervious to change. For over one hundred years, fossil fuels have produced countless products that form the basis for our economy and society, from syringes to cell phone cases, from aspirin to asphalt, from jet fuel to tennis shoes. Fossil fuels have also provided stable employment for generations of workers and a steady tax base for communities across the country. Sons, daughters, siblings, nieces, nephews, and cousins all work in the industry with the expectation that these jobs will continue for the next generation. Some will, but many will not.

The demand for fossil fuels is contracting, and so the industry is changing. Ford is spending almost $4 billion building electric vehicle facilities in Michigan, Ohio, and Missouri. Delta,

Southwest, and many other airline carriers have placed large orders for sustainable aviation fuel.[2] Volvo is set to offer only electric vehicles by 2030.[3] The company is also exploring using steel that has no fossil fuel footprint. Finally, General Motors will no longer be making internal combustion engines after 2035—another hit to demand for fossil fuels.[4] At the same time, oil lobbies are making the case that the industry is indispensable, all the while cutting staff, closing facilities, and moving to renewable fuels—all of which negatively impact our workers.

Our membership is more divided than the rest of the nation on the link between fossil fuels and climate. Some workers realize the need for change to protect the environment; others do not. What we tell these members is that whatever they believe about the climate, the fossil fuel industry is changing and we need to adapt with it. Talking about demand has allowed us to move the conversation within our membership beyond climate change and toward what we need to do to make sure our folks have good jobs moving forward.

Environmentalists, on the other hand, don't always take into account the wide-ranging effects of rapidly phasing out fossil fuels. For example, California recently announced plans to reduce crude oil production to 166,000 barrels a day by 2045.[5] Currently, my refinery alone produces 363,000 barrels a day. Refineries across the state produce around one million barrels a day.[6] The plan that's currently in place doesn't fully address how we would reduce production so dramatically or what the consequences of doing so would be—including the loss of jobs. The call to action from environmentalists has too often ignored the consequences for families and communities of reducing fossil fuel.

Over the last year and a half, labor and environmental groups in California have begun meeting to better understand each other's positions and to develop new platforms that take into account the needs of workers, communities, and the planet. As a result, we've begun to build alliances that can ensure that our communities continue to thrive as we transition away from fossil fuels. These kinds of alliances are crucial to ensuring that no one is left behind as we plan for a renewable future.

Together, we can escape the nightmare and win the world of our wildest dreams.

2

"OUR WORK IS WHAT MAKES THE FOOD SYSTEM GO"

AN INTERVIEW WITH EDGAR FRANKS

Jeff Ordower

Edgar Franks is the political and campaign director of Familias Unidas por la Justicia, an independent farmworkers' union in Washington State that was founded in 2013. Familias Unidas negotiates contracts with employers and builds farmworker cooperatives that engage workers in tending the land—with the larger goal of developing food sovereignty and mitigating against climate change.

How did Familias Unidas get started?

The union was started because workers wanted a contract where they got fair wages, where people weren't arbitrarily fired, where they had good housing, and where they had a say in their workplace. But we had to take on a corporation to do that. So, we had to organize. It's especially hard to organize when you're an immigrant: you don't have papers, you don't speak English. Some of our Indigenous members don't even speak very much Spanish. And we're in one of the poorest sectors in the whole country. So, there are all these things that make it hard. But the conditions are so undignified and there's only so much that one can take.

We originally started organizing in the summer of 2013. There was the blueberry harvest, and workers were working eight to ten hours a day, but because they are compensated by how many pounds they produce, they were not even making the state minimum wage. They were told that they had to pick faster or else they'd get fired. But there was not even enough fruit to pick. So a worker said, "Well, if you want us to work harder, then you have to pay us more." And everybody agreed with him. So, they walked off the job, just for that day.

How many folks walked off the job?

That was around three hundred workers. When everybody went out, the worker that had brought up the issue about getting paid more was told to go to the management office. When he showed up, they're like, "You're intimidating people. You're causing problems, so you have to leave right now." They gave him two hours to clean his cabin and leave. He was a fairly young guy. Him and his wife had just had a baby. Even if they wanted to leave, they couldn't because they had no money to travel across the state or to a different farm or to California.

So, he went around the labor camp and basically just said, "Hey, can you help me out with what's happening here?" The workers rallied around him. None of them went back to work until he got his job back. That was the incident that kicked everything off. That same day, workers came up with fourteen demands, and if they weren't met, then they were going to go on strike. It was about to be peak harvest, and the boss couldn't risk three hundred-some workers going out on strike because that would put their production for the whole season in jeopardy. So they agreed to a lot of the demands.

Then the company started going back on its word and pulled

out of those agreements. So the workers said, "Well, you know what? I thought we had something good. We had a paper that was signed by both sides agreeing to all this, but obviously the company's word is not good enough." So, I think everybody thought, "We'll just get a union and get a contract and get everything guaranteed. If they don't, then they'll have to go to court." Then we learned that you cannot call for a union vote right there and be recognized. We learned about the NLRB and NLRA, all these legal things that had happened before, all these laws that excluded farmworkers. Workers were like, "Well, that's not right. Why can't we be a union like everybody else?" Instead of getting discouraged, that just gave them more reason to organize, to fight back against bad laws that exclude farmworkers.

So it became bigger than just the union contract then. We started realizing the scope of what was happening and the opportunity that we had. That was in 2013. I think the goal was to have the company sit down and negotiate in good faith. But because of the history that they had, the only way that the company would listen to the workers was if the workers took their case to the public.

The company had its own label where they would sell their products—the raspberries, blueberries, blackberries, strawberries. So the workers started a boycott campaign to get all the company products off the shelves here at the local food markets and everything. However, later we found out that the company was also distributing through Driscoll's. Again, none of us knew what Driscoll's was. We were just like, "Oh, Driscoll's is just a label that the company is selling." Later, we found out how powerful Driscoll's is around the world, that it's one of the biggest distributing operations in the world for fruits. Even though Driscoll's says they don't own any farms, they do have a lot of say over what happens. We decided to boycott Driscoll's for three and a half years in the hope of putting so much economic and public pressure on

Driscoll's that they would sit down and negotiate with the farm where the workers were at. It actually did work! We went on this big tour and did public action campaigns all around the country and in Canada and Mexico. There were actions everywhere, and it became a huge international campaign.

The company started hurting economically and decided to sit down and negotiate. Well, first they allowed for a vote. We had to bring in an NLRB board member to supervise the election on September 12, 2016, and 80 percent of the workers voted in favor of the union. That's how the union actually became a union. Then we thought, "Oh, that was it." Again, all of this was brand new to us. We thought it was over. We didn't have to do anything. But you still have to negotiate a contract. That process was like a whole different thing. It's different than doing public actions and protesting. Now, you have to actually negotiate all the demands and everything that you are asking for, which is very difficult.

But luckily, the negotiation team from the union, which is all workers, put together one of the best contracts for farmworkers in the United States. It gets better every negotiation. That's just from something that we never intended to happen becoming something bigger—a model where workers can show workers that they can do a lot despite many of the obstacles. With organization and with support and solidarity, you can accomplish a lot.

How big is the union? How many members?

Right now, the union has about five hundred members who are under the collective bargaining agreement. We also have farmworker members who are not under the union contract, but because people need representation, they call us and we help them. They're not dues paying, but we still support them. So we have five hundred dues paying under a collective bargaining agreement, and another five hundred that are just members who we help whenever there's issues that they need help with. So, it's like around one thousand members all around the state.

What is the role of farmworkers in changing our food system? Farmworkers have firsthand experience of climate change, and this knowledge and experience can be one of the solutions to addressing the climate crisis. But farmworkers' knowledge is not taken into account because it's a difficult industry to organize, because of racism and xenophobia, and because of their class status. They're seen as just workers. But our work is what makes the food system go. Right now, our labor is building up a bad agricultural system, but what if we use our labor to do something good? That is why it is important to organize and build collective power to push back against the bad system—but also create space to really talk about how we transform it. It is good to have union contracts, but then, how do we go beyond union contracts and think about a world in which we're not beholden to companies or corporations anymore?

The co-ops happened almost by accident. I guess a little bit by accident. The workers were on tour to help get support for the boycott of Driscoll's in 2016. People would ask: "Is there any farm that you know of that treats their workers good and is doing the right thing?" Everybody was like, "Man, I can't think of one. In good conscience, I can't think of one place that is doing any of that." So the workers are like, "What if we do it? What if we show that there's a whole different way of producing food with workers, with what people want: organic, pesticide-free, good wages for workers, good treatment?" That was the beginning of the idea of workers having control of their labor with the values that workers wanted to see in farming.

Right now, the cooperative is located in Everson, Washington. It's sixty-five acres. It used to be a conventional raspberry field. But it's been five years already since the co-op has been established, and they've transformed that land. Before, when it was a conventional field, it was just getting sprayed with pesticides and it's monocrop agriculture, all for export pretty much. The co-op has kept some of the berries but is diversifying the use of the land. There's blueberries, raspberries, but there's also things that the community eats, so a lot of corn, a lot of tomatoes, chiles, basic food things. There's a greenhouse with cactus. There's this particular kind of squash that a lot of people from Oaxaca and Guerrero use in their food. It's called *chilacayote*. There's also chickens, so people can get eggs. There's goats for milk or for people who want to eat goat meat. So it's become a diversified farm now.

The goal is to have it made into an *ejido*. In the tradition of Mexico, of the communities where farmworkers come from, an ejido is a community-owned piece of land that can't be sold privately. In the ejido, the plan is to have housing for worker-owners, so they can just work and live and be there on the land with their families. That's still being developed. That's our model of a just transition. It gives people options of what it means to be a farmworker in the United States.

What have you learned?

There's got to be a process of consciousness-raising, of showing people that if you fight back and you organize, that it's possible to win. We might not always win, but we will be guaranteed to lose if we don't try anything. I think that consciousness-raising has been one of the things that the union wants to really emphasize, just to see workers themselves with power. We're a union; we have power, and if you want to organize, we want to support you no matter where you are in the state. Consciousness-raising has been one of the big goals of the union.

Another component is serving as a model of a just transition and what food sovereignty is. The agricultural system does need to be in a transition. It's got us to this point of being on the brink, where food is now treated as a commodity and not valued for what it is. It's the one thing that still connects us to our cultures and to who we are. It brings people together, and once you put a monetary value on it, it just becomes another thing that gets swallowed up by capitalism. We want food to be for people, not to be thought about as a commodity. This is who we are. Especially if our labor is being used to produce it, we should be part of the decision-making, of creating any system with that food.

Is there anything you want to say by way of closing?
Workers right now are in this position where there's opportunities for mass mobilizing and organizing to re-envision the labor movement for this next century or beyond. I think COVID really showed how little respect there is for all labor. That's why we're seeing a wave of strikes and organizing and unionizing campaigns. People saw that it's now or never. Right now is when we have to take a stand and really push back and protect the lives of regular workers and community people. The system that mistreats workers is the same system that's creating climate chaos. There's this moment of connections being made between workers and environmentalists to really change the system. That's been one of the exciting parts of organizing in this time.

3

PUSHING FOR A GREEN NEW DEAL FOR EDUCATION FROM BELOW

Todd E. Vachon

On August 23, 2005, teachers and students along the Gulf Coast made their way to school on what was ostensibly an ordinary Tuesday morning at the beginning of a new school year. Unbeknownst to most, a tropical depression was forming over the southeastern Bahamas, becoming Tropical Storm Katrina the next day. The storm, which tracked west from the Bahamas, gradually intensified. It made its initial landfall along the southeast Florida coast on August 25 as a Category 1 hurricane with 80 mph winds. After moving west across South Florida and into the very warm waters of the Gulf of Mexico, Katrina intensified again. On August 28, the storm attained Category 5 status, with 175 mph winds.

New Orleans mayor Ray Nagin ordered the first-ever mandatory evacuation of the city, calling Katrina "a storm that most of us have long feared."[1] Authorities used contraflow lane reversal on Interstates 10, 55, and 59 to speed up evacuations. However, many parishes were not able to provide sufficient transportation for citizens who did not have private means of evacuation, and fuel and rental cars were in short supply. The end result was that tens of thousands of residents were unable to evacuate, including some

30,000 who took refuge at the NFL Superdome stadium and another 25,000 who weathered the storm at the convention center.

When the storm hit the city, it unleashed a devastating amount of rainfall, causing the levees to fail. By August 30, the city was 80 percent underwater. Ultimately, the storm took the lives of more than 1,800 people, caused more than $160 billion in damage, reduced the population of New Orleans by 29 percent, and completely destroyed 110 of the city's 126 schools. The children who survived the storm—close to 400,000 students—were displaced to other states for the rest of the school year. Students who returned in the following years came back more than two years below grade level on average, some much more. By 2020, the city's population remained 20 percent below its pre-storm level. In response to these changes, the city terminated 7,500 unionized public school teachers and, in one of the most blatant acts of disaster capitalism, replaced the shattered public schools with charter schools.

While individual weather events like Katrina cannot be directly attributed to climate change, the likelihood, frequency, and intensity of hurricanes, floods, and wildfires are all increased by climate change. The impacts on students, teachers, and communities are overwhelming.[2] In 2021, school closures caused by extreme weather affected more than 1.1 million students.[3] The U.S. Government Accountability Office found that over one-half of public school districts—representing over two-thirds of all students across the country—are in counties that experienced presidentially declared major disasters from 2017 to 2019. The number continues to rise. Recent research in the journal *Scientific Reports* finds that school closures caused by extreme weather generate significant negative impacts on academic performance among students.[4] The connection between the increasing number of disaster-related school closures and lost learning is just one example of how climate change is affecting education.[5] The psychological and emotional toll of climate catastrophe on children is just beginning to be studied, but early findings also suggest increased mental health problems and conditions, including depression, anxiety, and general distress.[6]

As the stewards of the next generation, teachers are not standing idly by as increasingly unnatural disasters devastate schools and communities. Through their unions, many have taken both symbolic and concrete steps to confront the climate crisis. This article explores some of those actions and draws lessons for teacher and community organizers seeking to build a more just and sustainable education system and economy for the present and future generations. I examine what a union- and community-led Green New Deal (GND) for public schools and universities could look like, highlight some of the efforts currently underway at the local level, and outline the ingredients needed to make it a reality.

Teachers Resolve to Confront the Climate Crisis

In July 2020, the 1.8-million-member American Federation of Teachers (AFT) adopted a resolution in support of the Green New Deal, becoming the first and only national union affiliated with the AFL-CIO to do so. The resolution acknowledges that working people are facing dual crises of ecology and inequality—and that we cannot successfully address one crisis without addressing the other. The union resolved to support policies that enable local communities to develop, produce, and own renewable energy to reduce fossil fuel emissions "in accord with the latest climate science."[7] Noting that working families, frontline communities, communities of color, and low-income communities suffer disproportionately from environmental degradation and climate change, the resolution supports prioritizing investments in communities that have been "disproportionately impacted by pollution, high unemployment, poverty and environmental injustice."[8]

To address the job loss that will result from climate change impacts as well as the phase-out of fossil fuels, the resolution calls for "a just transition" that includes robust investments in union career opportunities to address the harms of both climate change and extreme inequality.

The GND resolution was introduced to the national convention by several local unions, which each drafted and adopted their own resolutions. Through the machinations of the union governance structure, the locals pushed these resolutions up through state federations and on to the national union, which compiled them into a single resolution to be considered by delegates. The delegates—made up of PK-12 teachers, health care workers, college and university professors, and school and college support staff—voted overwhelmingly to adopt the resolution.

Following a similar model, a small group of local unions advanced another climate resolution at the 2022 national convention, this time calling for divestment from fossil fuels. The resolution, which was adopted with broad support by the delegates, urges union retirement fund managers to divest their assets from "all corporations or other entities that extract, transport, trade or otherwise contribute to the production of coal, oil and gas" and to reinvest those funds in projects that "benefit displaced workers and frontline communities in the state or region of the given AFT members."[9]

In the time since these important resolutions passed, the national union has taken some small but important steps to promote a Green New Deal, including forming a national climate justice task force to oversee the union's climate work, including implementing the divestment plan. The union has also forged a powerful partnership with the United Auto Workers (UAW) to promote the purchase of union-made electric buses by school districts across the country. However, much like the impetus for the two resolutions, most of the climate justice work being done by teachers is happening at the local and state level—led by rank-and-file members through their local union climate justice committees and in partnership with community organizations.

A Green New Deal for Education

In the summer of 2021, New York representative Jamaal Bowman, a lifelong educator, introduced the Green New Deal for Public Schools Act in the U.S. House of Representatives. The national plan aims to invest $1.43 trillion over ten years in public schools and infrastructure to combat climate change by upgrading every public school building in the country, with a priority on addressing historical harms and inequities. If enacted, the legislation would fund 1.3 million jobs per year and eliminate 78 million metric tons of carbon dioxide annually. Given the current political gridlock in Washington, the prospect of the legislation passing in the near term is in doubt. But that does not mean a GND for education is impossible. Educators and school employees can push for a GND for their schools and universities from below—in fact, many already are.[10] Some of the key elements of a GND for education from below are investments in green school infrastructure, prioritization of environmental and climate justice, an open and democratic process for decision-making, and the creation of good union jobs in the process.

Creating Green Schools

There are more than one hundred thousand schools across the United States that together emit 78 million metric tons of carbon dioxide annually. Schools spend about $8 billion on energy per year—making it the second-biggest expense after personnel. On average, U.S. school buildings are fifty years old. They operate outdated and inefficient HVAC equipment and have poor insulation, and their electrical and plumbing systems are in desperate need of repair. The nation's 6,500 two- and four-year higher education institutions are no better off.[11] Creating green schools and universities can improve student health and learning, reduce greenhouse gas emissions, save school districts and taxpayers money, and create hundreds of thousands of local jobs in the process.

So what are the elements of healthy green schools?[12]

Green school projects include installing solar panels or other renewable energy sources, improving HVAC systems by installing heat pumps, making energy efficiency improvements to existing buildings (new doors, windows, and insulation), installing battery storage for onsite renewably generated electricity, creating microgrids that can support local resilience hubs during power outages, modernizing lighting to increase efficiency, switching from diesel to electric vehicle fleets, automating building energy systems (smart buildings), and creating more green spaces. New construction projects can also be designed using the latest technologies from the start to maximize energy efficiency and savings.[13]

The Inflation Reduction Act (IRA) of 2022 offers many incentives for local schools to make these upgrades. In particular, the AFT and other nonprofits lobbied for the inclusion of "direct pay" incentives, which allow tax-exempt entities such as local governments, school districts, universities, nonprofits, and unions to receive direct rebates, in lieu of tax credits, from the federal government to cover a significant percent of the cost of green projects.[14]

The IRA incentives are like grants, equal to at least 6 percent and up to 60 percent of any renewable energy project's cost. However, unlike regular grants, they are a matter of right—there is no competitive application process. If a school district makes an appropriate investment, the IRS will wire them money. The credits are applicable to the cost for fuel cells, solar systems, small windmills, qualified offshore wind, geothermal heat pumps, and energy storage. Projects that pay the local prevailing wage and hire apprentices from locally approved apprenticeship programs qualify for a 30 percent credit. Projects that meet the domestic content requirement earn an additional 10 percent credit. Projects in energy communities or low-income communities can earn up to an additional 10 percent credit each.

With direct pay, schools and universities can own their clean power and maximize their cost benefits in the long run by keeping 100 percent of the savings. The rebate can be used to pay off huge portions of the upgrade project immediately, and then the utility cost savings from self-generation of electricity can be used to pay off the balance.

The important thing to note about green school projects is that they must be initiated locally, through local budgeting processes, including bonding discussions, municipal capital budgets, and referenda. In other words, the incentives of the federal IRA can support thousands of local GND projects, but they must be initiated at the local level.

Education unions are strategically positioned to lead this effort, and in many cases, they already are. For example, the Seattle Education Association—together with the International Brotherhood of Electrical Workers (IBEW)—passed a resolution at their labor council to pressure the school board to fund green school upgrades with project labor agreements to ensure the work was done by unionized tradespersons.[15] Teachers, parents, and students all testified at board of education meetings and won an additional $19 million for green school upgrades. Feeling the power of their organizing success, the coalition then successfully fought for $100 million to update community centers in the city and are now pressing for a study to estimate the cost for a complete overhaul of all school buildings. Similarly, the Boston Teachers Union has begun discussions with the city's school board and City Hall about recently elected mayor Michelle Wu's plan for a GND for Boston schools. In this way, local education unions around the country, in partnership with community allies, can lead the process of greening our nation's schools now.

Advancing Climate and Environmental Justice

A second key component of a GND for education is the prioritization of environmental and climate justice to secure an equitable distribution of environmental burdens and benefits. For example, lead in public water supplies is a tremendous health hazard to students and residents in frontline communities. Workers and community activists can advocate for the repair or replacement of poisoned water pipelines and demand the cleanup of the groundwater and aquifers that feed those pipelines. Education unions and community members can also demand the electrification of vehicles to reduce student and worker exposure to asthma-causing particulate matter pollution as well as to reduce greenhouse gas emissions.

Creating more green spaces in urban communities and constructing bike paths or walking trails can reduce auto traffic around urban schools and universities that already suffer disproportionately from air pollution. To address historical inequities in the job market, unions and community partners can demand that a percentage of hiring for green projects be from the local community where the facilities are located, an approach championed by Jobs to Move America. Toward this goal, a union preapprenticeship program could be established to create a pipeline for local residents to enter careers in the green economy, including the construction trades.

Community-owned solar, microgrid, battery storage, and resilience hubs are key ingredients to equitable climate resilience. When the grid goes down, PK-12 schools, as local anchor institutions, can provide a safe space for the provision of potable water, electricity for charging medical and communication devices, refrigeration for medications, and other vital services needed to save lives during climate catastrophes, such as hurricanes. The same is true of public universities. In fact, the Rutgers AAUP-AFT, along with community partners in Camden, has been building support for community-owned solar and demanding that the university create resilience hubs to serve host communities.

Creating Good Union Jobs and Career Pathways

Climate equity also means ensuring that the new jobs created are good jobs, providing opportunities not only for workers from historically marginalized communities but also for those displaced from the fossil fuel industry.

As noted above, the IRA promotes strong labor standards by providing additional incentives for projects that offer prevailing wages and utilize apprentices from qualified apprenticeship programs. By taking wages out of competition in the bidding process, high road union employers have a better chance of securing contracts to retrofit old schools or build new green schools. Apprenticeships create a career pathway into good-paying blue-collar jobs without the burden of debt that most working-class students accrue pursuing college degrees. Sourcing building materials from domestic manufacturers also helps to support local manufacturing job opportunities. Additionally, project-labor agreements and community-labor agreements are proven ways to enshrine strong labor standards and hiring protocols into new developments before the projects even begin. Education unions, in partnership with local construction and manufacturing unions and other community partners, can use all these tools to ensure good jobs are created through their GND plans.

Ensuring an Open and Democratic Process

Social and economic justice campaigners operate under the simple principle that "nothing about us, without us, is for us."[16] It means that decisions that significantly affect people's lives cannot be fair and just if those people are not included in the decision-making process. Ensuring meaningful participation by all stakeholders is essential not just for moral but also for practical reasons. Comprehensive climate protection measures have proved so elusive in large part because of the lack of a broad coalition to support them.

How We Win It—from Below

Form Local Union Climate Justice Committees

Most climate action around education is being led by members of unions that have adopted climate resolutions and have formed local union climate justice committees to advance work on the issue. Forming a climate justice committee does two things. First, it ensures that the issue of climate justice remains on the union's agenda. Second, it creates a space for interested members to engage with the issue within their union and helps to drive the union's climate work at the grassroots level. Local unions that do not yet have a climate justice committee should form one. The more local climate justice committees that exist, the more local unions there will be pushing a climate justice agenda within their school district or university.

Build Transformative Partnerships to Win Transformative Changes

Building the broad base of support that's needed to win a GND locally or nationally means building strong and trusting relationships among multiple stakeholders. In the pursuit of climate justice, education unions must forge partnerships with local environmentalists, environmental and climate justice groups, student and parent organizations, and other unions in different industries.

Many education unions have already been engaging in this work, including the United Teachers Los Angeles (UTLA), which has more than a dozen community partners with whom they are making climate justice demands at school board meetings and in bargaining. Community partners include the Alliance of Californians for Community Empowerment, Angelenos for Green Schools, Climate Reality Project LA, Democratic Socialists of America Los Angeles, Los Angeles Alliance for a New Economy, LA Compost, Northeast Los Angeles Climate Collective, and Ready to Help Mutual Aid Network. These labor-community partnerships became a priority for the union following a leadership change in 2014 that reoriented the union toward an organizing, democratic, racial justice, class struggle model of unionism.

The Seattle Education Association has also built strong ties with parents, students, and local building trades unions to increase funding for green school investments in the city. Seeing energy efficiency upgrades as an opportunity to not only make healthier green schools but also create good local construction jobs, the teachers—along with the local chapter of 350.org—partnered with the IBEW to win support from other unions. They began by introducing and winning a resolution at the local central labor council that urged unions and community partners to lobby together and make public comments at meetings to increase the budget for green school upgrades. The effort was successful in large part because the group of stakeholders making the demands cut across the bases of many elected officials. The political influence of the local construction unions also played a large role.

Milwaukee teachers have similarly partnered with community allies to promote sustainability projects in Milwaukee-area schools, including schoolyard redevelopment projects. The work really began after the adoption of a climate justice resolution titled Our House Is on Fire by the Board of School Directors of the Milwaukee Public Schools in 2020.[17] The resolution calls for Milwaukee Public Schools to integrate climate justice curriculum and sustainability practices into all aspects of its work and functions. The work is inclusive of multiple stakeholders, including schools, government agencies, and community organizations, such as the Milwaukee Area Technical College, University of Wisconsin-Milwaukee, Milwaukee Metropolitan Sewerage District, Reflo, Green and Healthy Schools Wisconsin, City of Milwaukee Environmental Collaboration Office, Department of Natural Resources, and various gardening and agricultural organizations.

Even in red states, there are opportunities to forge powerful relationships with students and parents to fight for and win the healthy green schools our children and communities deserve. In Texas, where teachers do not have the right to bargain collectively with their employers, local union leaders in Austin, Dallas, and Houston have been forging powerful relationships with their communities and running and electing candidates for local boards of education to ensure the safe return of students to schools after the initial wave of COVID-19. These partnerships have flourished, and the scope of demands is expanding to include other issues, such as climate justice. For example, in Austin a coalition made up of Texas Climate Jobs Action, Education Austin, several building trades unions, and the Austin DSA organized in support of the Better Builder Program, which would establish stronger labor standards for construction work on schools in the city. Following petitions and public testimony before the school board, the program passed unanimously, raising the wage floor and benefits for thousands of construction jobs in Austin. The coalition also helped to elect a union majority to the school board where decisions about healthy green schools will be made in the coming years.

In response to collaborative campaigns from students, parents, and climate-focused organizers, a growing number of school boards have passed clean energy resolutions, including the Los Angeles Unified School District (100 percent clean, renewable electricity by 2030), Miami-Dade County Public Schools (100 percent clean energy by 2030), Salt Lake City School District

(100 percent clean electricity by 2030 and 100 percent carbon neutral for all operations by 2040), San Diego Unified School District (100 percent clean energy by 2035), and Seattle Public Schools (100 percent clean, renewable energy by 2040).

Bargain for the Common Good

The key to strong partnerships is trust, and building trust takes time. Local unions pursuing climate justice that do not already have existing relationships with community partners should begin to open up dialogue like we have seen in Austin, Los Angeles, Milwaukee, Seattle, and elsewhere. And while forming coalitions can lead to tangible gains, to win truly transformative changes we need transformative coalitions that involve radical power sharing and democracy, as is the case in Bargaining for the Common Good (BCG) campaigns.

BCG is an innovative way of building community-labor alignments to jointly shape bargaining campaigns that advance the mutual interests of workers and communities alike. It developed over the last decade out of the struggles of teachers and public employees in St. Paul, Chicago, Seattle, San Diego, and Los Angeles, where unions partnered with their community allies to advance a common agenda through direct action protests—including strikes—and contract campaigns that targeted the power structures of their communities.

At their heart, BCG campaigns seek to use union bargaining to increase employer investment in underserved communities and to confront existing inequalities. They do so by expanding the scope of bargaining beyond wages, hours, and working conditions to include demands that address structural issues around education, racial justice, public services, immigration, finance, housing, and privatization. Successful BCG campaigns engage the members of both unions and community organizations, creating opportunities for deep relationship building and joint visioning.

BCG campaigns are perhaps best suited to taking on the existential crisis of human-caused climate change—which intersects with and exacerbates all of the other issues facing communities. Some unions have already begun to incorporate climate justice demands into a BCG framework. For example, on February 27, 2020, more than four thousand members of the Service Employees International Union (SEIU) Local 26 in Minneapolis went on strike demanding not only improvements in wages and working conditions but also action on climate change. Working with partners from Minnesota Youth Climate Strike, Environment Minnesota, the 100% Campaign, MN350, the Sierra Club North Star Chapter, and the Minnesota Black, Indigenous, and People of Color Climate and Environmental Justice Table, the union presented a set of climate justice demands that included the creation of an owner and community green table; closure of the Hennepin Energy Recovery Center incinerator, a major source of both greenhouse gasses (GHGs) and air pollution that harms nearby communities of color; and adoption of the union's proposed green training program.

Some education unions at large public and private institutions are also working in partnership with local environmental organizations and climate justice activists to develop bargaining demands that promote climate justice. For example, UTLA, in partnership with a coalition of community organizations called Reclaim Our Schools LA, engaged parents and students to develop the Beyond Recovery platform as a transformative plan for students and schools. The platform proposes an entire article be added to the union contract stipulating the elements of healthy, green public schools. Some demands include reducing GHG emissions, making buildings more energy efficient, installing and owning solar, divesting public pensions and endowments from fossil fuel companies, and moving those funds into socially responsible investments. Other demands include the expansion of public transportation options, including the free provision of mass transit to students and employees, and monetary or other incentives for workers who walk, bike, or use public transportation to commute to and from school.

The coalition has urged the superintendent of the Los Angeles Unified School District to bargain over all the proposals submitted by UTLA, including those that address the climate crisis. Thus far, the district has rejected all the common good proposals. However, the same was true of similar demands the union and partners made during the 2018 contract campaign, which led to an eight-day strike of thirty thousand teachers in LA. In the end, the teachers won not only wage increases, a reduction in class sizes, and the establishment of thirty community schools but also a host of other common good demands, including increased green space on school grounds. Important to note, these wins were only possible because the union had successfully organized a supermajority of members to support a strike action.

Public school teachers in Chicago are also advancing common good demands. Along with community partners, the Chicago Teachers Union (CTU) is developing a climate change curriculum with a strong focus on equity and justice—and bringing this demand into contract bargaining. The curriculum program, called the Climate Justice Education Project, seeks to win environmental justice and clean energy transformations in schools by developing educational and advocacy opportunities for Chicago educators, students, and communities. The project supports teachers in designing and implementing climate justice lessons to their students. By teaching fellow educators and students how climate change works, how to investigate climate problems and solutions, and how to get involved in civic action for climate justice, the project will empower students and educators to demand green policy solutions that transform the physical infrastructure and everyday practices of their schools and communities.

As these examples illustrate, a BCG approach starts with unions and local community groups working together to develop and articulate a set of demands that serve the interests of workers and the communities where they live and work. To succeed, they require power building.

Make a Table and Secure Seats for Labor and Community

Without clearly defined targets and an enforcement mechanism, green school plans are simply promises that can be broken when economic or political winds shift. To ensure that educational institutions follow through on their goals, some unions have demanded the formation of joint labor-management-community committees on GHG emissions reduction. The AAUP-AFT at Rutgers University, for example, made such a demand during postdoc negotiations in 2018, and although they did not win during bargaining, they did win four years later after a number of protest actions. The committee currently meets quarterly to discuss ideas and concerns and to assess progress on the university's climate goals.

In workplaces without established climate goals, such committees could assess the employer's emissions profiles and help to develop climate action plans to reduce GHG emissions and promote climate justice, including the creation of resilience hubs and career opportunities for local community members. Instead of relying on politicians who may be too fearful to establish enforceable targets or take bold action, workers and community partners can persuade or, if need be, force their employers to do so.

Coordinate Efforts Nationally

The impact of local efforts is amplified when they are happening in multiple locations at the same time. One way to help coordinate local efforts is to create a climate caucus within each of the major national unions. Such an effort is currently underway within the

AFT. The goal of the caucus is to help advance the work of the national climate task force, offer creative ideas and solutions, and, if needed, push from below for more bold action.

A network has also been formed to coordinate efforts across the two largest education unions: the National Education Association (NEA) and the AFT. The Educators Climate Action Network (ECAN) emerged after conversations by education union activists at the Labor Notes conference as well as at the national AFT conference in the summer of 2022. The network of over one hundred union educators from across the country is convened monthly by the Labor Network for Sustainability (LNS) and is open to all education union members interested in tackling climate change and promoting climate justice.[18]

Organize, Educate, and Agitate

Together with parents and community members, education unions and students can build a powerful force for a GND for education to advance climate justice and address not only the climate crisis but also the inequality crisis. As with all major societal change, doing so begins with organizing and building power and then exerting influence on decision-makers to advance an agenda that promotes equity. Many education unions are already doing this work, but the potential for transformative change has only just begun to be tapped. Starting at the local level and coordinating nationally, education is a prime space for advancing a GND from below. Let's get to work.

4

CARE WORK IS CENTRAL TO A JUST TRANSITION

Batul Hassan

When wildfires raged across Northern California in 2017, union nurses were among the first responders.[1] Smoke filled the Kaiser hospital in Santa Rosa while fires raged within view. It took hours to complete the evacuation. Filling the gaps created by burn unit capacity reductions over the preceding decades, nurses in hospitals across the region took on the work of addressing the immediate needs of patients affected by the fires. In many cases, nurses forced to evacuate their own homes returned to work at hospitals that remained open because they had nowhere else to go. "This level of disaster is a new one for us," said Beatrice Immoos, an RN working at Kaiser San Rafael at the time. "It was very emotional, but there was a lot of resolve. Every day, nurses are always working with the common goal of taking care of our patients, and in a disaster, it's just even more hands on deck of working to get them the best treatment."[2]

Care workers like nurses are already frontline workers in the climate crisis—caring for people forced to seek care because of air pollution, lack of access to clean water, increasing levels of climate anxiety, heat-related health threats, and emergency medical needs caused by power outages, not to mention longer-term environmental health concerns. As the climate crisis intensifies, work like theirs will become only more essential—and will require greater power in the workplace to ensure the labor undertaken is safe and sustainable.

Care workers have the potential to form a coalition with the power to win better working conditions and to demand the transformative common good changes their communities need to weather the effects of a changing climate. Too often, however, care workers are left out of broader conversations about labor's role in building a decarbonized economy. Typically, those conversations center on energy and manufacturing workers as the front line of a just transition from fossil fuels. But much of the labor of building a decarbonized world rests on an equally essential workforce: care workers. The fastest-growing workforce in the United States, care workers are strategically positioned—in homes, schools, childcare centers, nursing homes, and clinics—to bring a renewable economy into being. To take on the crisis at the scale it requires, labor and climate movements should invest in organizing care workers.

The introduction of the Green New Deal (GND) in 2019 laid a path for a worker-led transition to a post-carbon world. The goal of the GND is to expand and mobilize state capacity to respond to the climate crisis in a way that creates new jobs and delivers benefits directly to communities, particularly frontline and BIPOC communities, which have borne the worst burdens of the extractive economy. These goals are not just morally good or just—they are also necessary for building a constituency that is broad and powerful enough to take on the climate crisis at scale. More than a piece of legislation, the GND offers a framework for thinking about how to address labor and environmental concerns simultaneously. Since the initial introduction of the legislation in 2019, others have built on it to develop new GND frameworks, such as a Green New Deal for Transportation, which would expand and electrify public transit, and a Green New Deal for K-12 Public

Schools, which would invest in retrofitting public school buildings and expand social services for students.[3] By bringing solutions for social and environmental crises under the same umbrella, the GND framework offers a tool for workers to strategize around a vision that not only urgently decarbonizes current energy use and fossil-fuel-driven infrastructure but also socializes ways of living and working within systems of production, consumption, and care.[4]

Organized care workforces are central to this vision. As the country's largest sector, care work—in homes, hospitals, schools, and childcare centers—represents one in seven jobs nationwide, with a higher share in deindustrialized centers. Home health aides alone represent one of the fastest-growing jobs in the country.[5] The rapid growth of the commodified care industry in the United States shows how interlinked care workers and their communities are—even as those links are increasingly filtered through a profit-driven private-care sector. Our communities' dependence on care workers is only growing. As increasing climate disasters deepen existing vulnerabilities and expose new ones, care workers—whether paid or unpaid—will be the ones to help frontline communities weather disasters.[6]

Despite their central role in our shared survival, care workers face large barriers to exercising their full power. The sector is stratified by race, with people of color performing more of the lower-paid work—like that of home health aides. Rich countries like the United States also rely heavily on migrant labor in the care workforce.[7] Even as the need for this work grows, workers in various parts of the sector face low (and lowering) wages and precarious working conditions, including dangerous levels of understaffing and unpredictable scheduling.[8] Take childcare, for example. In 2020, the average U.S. childcare worker earned less than $13 per hour, with few or no benefits.[9] People seeking childcare, meanwhile, often find it increasingly difficult to find it or afford it.

From February 2020 through December 2022, the total number of childcare workers in the United States dropped by ninety thousand people.[10] Half of the U.S. population were already living in regions defined as childcare deserts, or where fewer childcare spots are available than the number of people seeking them, a problem that is only getting worse.[11] Childcare is one of the largest expenses for working people with children. Average national costs exceed $10,000 per year—and in many places, the cost is much higher.[12] The availability and affordability of childcare has effects that ripple through multiple lives: in November 2021, nearly half of people with a child in their household pointed to childcare issues as a reason for quitting their job.[13]

Organized care workers are thus well positioned to turn the exhausting everyday burdens borne by working-class people into common, concrete, justice-oriented demands for change. Such demands can unite workers to take shorter-term agendas of traditional collective bargaining—which remain urgently necessary—and transform them into the multi-scalar agenda that is needed to take on the climate crisis at scale. Teachers are already doing this work, using the Bargaining for the Common Good framework to wage contract campaigns that move beyond mandatory bread-and-butter subjects of bargaining to address community-level needs and take on the root causes of problems in their classrooms. Their organizing provides a road map for how we can build labor-climate power.

In California, United Teachers Los Angeles (UTLA) has bargained on everything from small classroom sizes to expanded health services for students to access to green spaces. UTLA has been especially successful in centering racial justice demands. Los Angeles Unified School District is composed of 88 percent students of color and more Black students than any other district in California.[14] In response to the 2020 uprisings against police brutality, union teachers endorsed divesting school funds from the LA School Police Department and investing them in support services for students. Eventually, a coalition of teachers, students, parents, and community organizations—including Black Lives Matter LA—won the divestment of $25 million from one of the largest school district police forces in the United States as well as additional funding to increase supportive staffing and the implementation of a Black Student Achievement steering committee.[15] Using a similar model, the union has begun to pressure the district to bargain over climate justice proposals as part of its Beyond Recovery platform.[16] The union's proposals for healthy, green public schools include green spaces for students, funding for community gardens, a new electric bus fleet, strategic plans for "installation of solar panels, use of school land for collection of clean water, creation of schools as cooling zones," the transformation of schools into resiliency centers during climate emergencies, and healthy, sustainable cafeteria food options to address food insecurity and improve student nutrition.[17] As of December 2020, bargaining continues—with union teachers leading the way for a just future for the district in which they work.

UTLA is not alone in fighting for—and winning—climate justice demands. Teachers across the country have shown the power that can be built by deeply engaging a broad coalition of workers, community members, and community organizations. The Chicago Teachers Union (CTU-AFT Local 1), for example, has successfully bargained for free school transportation for low-income students, a demand that not only allows students to get to school without worrying about cost but also reduces the number of personal vehicles required to meet daily transportation needs. The transportation sector represents the largest sector of U.S. greenhouse gas (GHG) emissions and requires rapid decarbonization to meet climate goals that would keep the United States on track to meet global targets for a 1.5-degree limit to global warming.[18] While relatively minor increases in public transit use alone are not sufficient to hasten the decarbonization of the sector, increased ridership helps justify increased funding for electrified public transit options. Opening the topic as a workplace benefit can also help build solidarity along supply chains. By pushing for access to decarbonized transit systems or electric school buses, for example, workforces can also demand their employer purchase from union manufacturers and employ union drivers.

In 2018, the statewide Oklahoma Education Association campaigned for legislation that would translate a higher tax on fossil fuels into raises for public school teachers, with language demanding that "the state institute a higher tax on oil production, gas production, and motor fuels to fund public education."[19] Climate justice campaigns can also find common cause among workers from various sectors. The Seattle Education Association, for example, has partnered with building trades representatives and the local labor council on a campaign to invest $19 million over three years to solarize district public school buildings. The Rhode Island Climate Jobs campaign also brings a broad coalition of teachers, trades, and community groups together to fight for zero-emission school systems.

Unlike workers in many other industries, teachers are well positioned to fight on common good demands because their work grounds them in community and facilitates the development of deep relationships with students, parents, and other workers. Through decades of struggle and anti-labor attacks, public school teachers have managed to maintain union density and build the kinds of organizing structures that position them to lead transformative campaigns. As a result, unions like UTLA and CTU are strike ready—giving them the power needed to make viable strike threats to back up their common good demands.

Other care workers can apply these lessons to their organizing. Union nurses, for example, are well positioned to make a clear and urgent case for the need to upgrade hospital energy systems and invest in greener, healthier facilities and supply chains. The climate crisis and current land use practices will only increase our vulnerability to global pandemics like COVID-19, intensifying the stress on nurses and other care workers, such as nursing home and childcare workers.[20] At the height of COVID, too many care workers spent their shifts in high-risk settings, often without PPE or testing standards in place, and far too many lost their lives to make a living. Unions often made the difference. A study comparing union and non-union nursing homes in New Jersey showed lower rates of the virus among both workers and residents in unionized care homes: resident COVID-19 death rates were 10.8 percent lower and worker infection rates were 6.8 percent lower in union workforces compared with non-union nursing home workforces.[21] The difference between a protected, empowered care workforce and an unorganized, divided one—for both care workers and care receivers—increasingly represents the difference between life and death in the age of climate crisis.

As the environmental health burdens of the climate crisis worsen, health workforces and the communities they serve must be prepared for a range of emergencies, from deteriorating air quality to reduced food availability to heat waves, droughts, extreme storms, and flooding.[22] They must also be powerful enough to ensure employers and policymakers are compelled to protect vulnerable workers and community members. Community-based struggles for improved public health have historically been taken up by labor unions, which have served as community sites of health and wellness education, information, and improvement. Existing unions can serve as a protective factor in health emergencies for both workers and recipients of care.[23]

Other kinds of workers have shown the possibilities of organizing around common good climate platforms. In Minnesota, for example, janitorial workers staged what *Labor Notes* termed the "first union climate strike" in early 2020 using common good bargaining tactics to reduce carbon emissions at their workplace.[24] The four thousand janitors, represented by the Service Employees International Union (SEIU) Local 26, won a collective bargaining agreement with provisions to reduce waste and lower water and energy consumption, with a fund to be jointly developed with workers—in addition to wage increases, paid sick days, health benefit improvements, and protections against discrimination and harassment.[25] These workers demonstrated how the benefits of labor organizing for climate justice can go hand in hand with concrete, material improvements to working and living conditions. For caring professions, the lesson applies even more broadly: organizing care workforces for climate justice improves not only working conditions but living conditions for everyone who relies on the labor they provide.

As the climate crisis intensifies, it can be difficult to imagine the transformation from a fossil-fueled world to a regenerative one. A powerful care workforce—with the strength to leverage its own labor in coalition with communities for common good climate goals—is necessary to ensure a future in which the world is not only decarbonized but livable for all people. Jane McAlevey trains rank-and-file organizers to start any campaign by raising the expectations of their coworkers, slowly and steadily exercising the power of their workforce through structure tests, and eventually wielding that strength to win their demands. These same lessons can be applied to climate organizing among care workers. Labor-climate coalitions have identified their shared enemy: capitalist bosses who profit from the destruction of the planet and the impoverishment of its people. Acknowledging the power and necessity of an organized care workforce to ensure our common flourishing must be a central component of the struggle to fight these bosses as we build a healthy, just, decarbonized world.

5

YOUNG WORKERS CAN BRIDGE THE LABOR AND CLIMATE MOVEMENTS

Maria Brescia-Weiler and Liz Ratzloff

When Kathleen Rose was in her early twenties, she thought the best way to change the world was to work for a nonprofit. She was concerned about the environment and social issues, so she got a job at a local nonprofit working on sustainable agriculture. She quickly noticed that she was spending more time and energy negotiating funders' priorities—work that could be exhausting, frustrating, and demoralizing—than performing the actual work she had hoped to do. The need for social change was becoming only more urgent, but she was increasingly convinced that her work at the nonprofit wasn't the way to do it.[1]

Kathleen was "in the middle of a 'What am I going to do with my life?' moment" when she learned about the apprenticeship program for the International Union of Operating Engineers, a trade union for heavy equipment operators. After four years in the earn-as-you-learn apprenticeship program, she had full-time union work with benefits as an operating engineer. She was a member of her local Central Labor Council and an active participant in her union. She had also started to form strong bonds with other women in the trades—bonds that would become the foundation for her labor-climate organizing efforts. While Kathleen had felt stagnant and isolated working to fight the climate crisis in the nonprofit industry, her union work gave her the opportunity to educate herself about sustainable infrastructure needs and a home base in the working class from which to advocate for solutions to meet these needs. Kathleen is still on her way to figuring out how to move the building trades to become vocal and strategic advocates for climate justice. The powerful set of unions representing construction workers has often opposed any climate action. Still, Kathleen's sense of possibility is unwavering. "We just kind of have to be organized into a fighting union that can stand for something broad and visionary. And then also just kind of understand that all of our lives get harder and shittier as climate change progresses at work and at home."

Kathleen is just one of many young workers who want to use their position at the grassroots of the labor movement to build an economy that allows them to make a living on a living planet. This new generation of workers—who grew up in the shadow of the climate crisis and growing economic inequality—see the connections among environmental, economic, and social justice issues. Their leadership is our best chance at building a strong labor-climate movement.

Young workers have already demonstrated leadership on social and economic justice issues. From school climate strikes to nationwide protests against police brutality to recent union drives among the young workforces of Starbucks and Amazon, these workers are actively engaged in political work. But labor has been slow to capture the energy young workers can bring to the movement. In 2009, the AFL-CIO passed Resolution 55: In Support of AFL-CIO Programs for Young Workers, acknowledging the need to engage young workers, but a decade later, very few labor-affiliated young worker programs exist. The few that do—such as the Reach out and Engage Next-gen Electrical Workers (RENEW) program of the International Brotherhood of Electrical Workers (IBEW) and the NextGen programs of the Communications Workers of America (CWA) and United Steelworkers (USW)—are largely under-resourced or deprioritized. If the labor movement doesn't begin to invest in young workers, there is little chance that we will build the power needed to secure an ecologically sustainable and economically just future. Understanding the perspective of young workers is a crucial first step in bringing these workers into the labor movement.

The Young Worker Listening Project

The Labor Network for Sustainability (LNS) engages workers and communities in transitioning to a society that is ecologically sustainable and economically just. For more than a dozen years, LNS has worked to foster deep relationships between the labor and climate movements so that they can become key partners in confronting the climate crisis. There is considerably more work to do. Neither organized labor nor the climate movement has built the power or the relationships necessary to win transformative change together.

In 2019, LNS launched the Young Worker Listening Project to better understand the experience and perspective of young workers in regard to their workplaces, the climate crisis, and the labor and climate movements. Through this process, we surveyed nearly four hundred young workers and conducted in-depth interviews with seventy of them. We pulled from the network LNS has already built through years of work at the intersection of labor and the environment. We used our newsletter, social media, targeted email, and one-to-one outreach to connect with the leadership of union locals who were interested in understanding their young members better and rank-and-file workers who were looking for space to have these conversations. Our sample included interviewees from twenty-four states and the District of Columbia, including forty-four union members and five union staff members (two of whom are also union members) from more than twenty unions. Key sectors included the building trades, nonprofits, education, and government—industries that are central to discussions about climate change either because they play a key role in responding to it or because their work is threatened by it. While we did not ask every interviewee about their racial identity, we know that our sample reflects historical gaps in organized labor and the mainstream U.S. environmental movement in that the workers we interviewed were overwhelmingly white. Notably, we did not reach key sectors that employ disproportionately young workers of color, including farmworkers, restaurant and service workers, and warehouse and logistics workers.

Interviewees ranged from activists who had formed climate organizations as teenagers or held union offices to workers who are frustrated by our inhumane economy and curious about what it might mean to engage in labor or climate activism. Still, we found a lot of alignment in how young workers are thinking about the intersection of work and climate change as well as what they need out of their jobs, unions, communities, and government. We also learned a tremendous amount about how responding to these needs can lead to building a strong labor-climate movement. Several important themes emerged through our conversations.

We have no choice but to respond to climate change and economic and social inequality.

The young workers we spoke to agreed: the question is not if but how every workplace, institution, and individual should respond to climate change. Economic and environmental uncertainty have been the backdrop of young workers' lives, and they have an intuitive understanding of how deeply linked these issues are to each other as well as to issues of racial and gender inequality, militarism, and imperialism. As an electrical worker in Washington

State articulated, "No one thinks anything's OK. Everyone under 40 I know thinks we need significant change. Which I think is why the 18 to 35 demographic [is so pro-union] because I think we've all seen, like, well, the people who told us unions are bad were wrong about everything else."

Nearly all interview participants described how climate change has affected their work. Many workers spoke about the effects of extreme weather on their health and safety. As a postal worker from Oregon told us: "We have post offices being engulfed in flames, so obviously, those people can't get to work. When the smoke is so bad in the air that it's difficult to breathe, it's difficult to be at work." Other workers told us about the lack of adequate accommodations to deal with disasters—whether they were struggling to get to work in the face of extreme weather or working from home hundreds of miles from an employer who had no idea what was happening in their community.

Young workers also described the deep impact of the climate crisis on their personal lives. Many described intense feelings of anxiety, depression, and grief. A private sector sustainability manager reflected on the environmental degradation they had witnessed in Colorado: "I am deeply scared. If I'm not scared, I'm angry. That's pretty much where I'm at and that really fuels most of what I do in almost every regard." Many young workers reported that their lived experiences with climate change have driven them to take climate action. Others told us that climate change had caused them to think differently about what kind of work they wanted to do.

There are no inherently virtuous jobs, and all jobs can be good jobs.

The young workers we spoke to articulated the value of all kinds of work, the need to organize for better working conditions across sectors, and the importance of building solidarity among different kinds of workers. As an electrical worker in North Carolina shared: "I feel like trade unions are very disconnected from the plight of the working class. They don't relate to Fight for 15 because entry-level starting out for us is $15, so it's like, 'There's no way that somebody that's flipping burgers should make $15; I make $15.' It's like, maybe you should make more money." Like many of the young workers we spoke to, this worker recognized that there can be as much dignity in service work as there is in the trades—and that the survival of the labor movement depends on building worker power across industries in an economy that is increasingly dependent on overworked and underpaid workers in the service, care, and logistics industries.

Several workers we interviewed also described seeking out work they hoped would create some social good or political change, only to find that the working conditions made it impossible to carry out the mission of their organization. They recognized a need for better wages, more time off, greater job security, and more equitable workplaces. A nonprofit worker in California told us, "I think we need to get rid of the idea that work will love us and that this is our family and that this is not really a job because we're passionate about it. No, actually, if you need to work for a wage, you're a worker." Across industries, many young workers rejected the idea of work as a family or a calling. Their deepest loyalties are not to their jobs but to their coworkers, and they are thinking critically about how they can make their jobs more sustainable—for themselves and the environment. A refinery worker in California shared, "I believe if you find something that you really love to do and you see there's a problem with it, then most definitely you can try and fix that situation. But the jobs that I had weren't really career jobs—like I used to do security, I used to do Amazon, and I just feel like there was no room for growth. But this job that I'm at right now, there's definitely room for growth. So I am making pretty good money, but if I were to stay here even longer, I feel like most definitely I can try and progress in this opportunity that I have to make things better."

Our workplaces are central to the solution, and the solution is bigger than our workplaces.

Many of the workers we interviewed argued that workers must be at the center of the movement for climate justice. As an operating engineer in Massachusetts who has constructed pipelines and participated in environmental activism argued, "We're the ones that understand how the weather works and understand what finite resources really mean, and just how things are put together and what it takes to take care of people." Building solutions to the climate crisis requires the knowledge of workers—who are intimately familiar with the impacts of climate change and the tools we have to respond to it.

But the workers we interviewed often had trouble articulating specific ways to push their unions and workplaces to take action on climate, particularly in the face of actively uninterested bosses and leaders, limited capacity, and so many immediate crises. A public sector worker in California described their concern about climate and how it impacts their work in water management, concluding, "I think everybody's thinking about it in terms of their life, but I don't think anybody's thinking about it in terms of work."

Many young workers have keen insights into how their working conditions impact their capacity to effect change beyond the workplace. An electrician in California shared that union membership has allowed them more financial security than they've ever had, adding: "It's also cut my work hours . . . and that's opened up a lot more time for me to spend with my kids and my wife and also for me to do recreational things, like going to meetings, and now I can be more involved just overall with our union." Similarly, a private sector worker in North Carolina told us, "Taking a job where I have financial security allows me not just to meet my own needs, but also allows me to be more engaged in the kinds of redistribution efforts that I want to be involved in."

Almost all the young workers we spoke to maintained hope in the democratic infrastructure of unions as a venue for effecting change. As an electrical worker in Washington said, "The people we elect to run the union, they just work for us. They don't have any power that we the membership don't have." Many young workers we spoke to are pursuing economic security, but they also want a sense of community and belonging as well as the ability to effect change and engage in grassroots democracy. They recognize unions as one of the few avenues for attaining all of this at once.

A grassroots movement is critical for effecting change. It is also a source of joy and community.

Many participants talked about the enormous potential of effecting change when working people get together and demand it. They recognized that we are past the point of simply recycling or walking to work: responding to climate change requires major systemic change. We need worker activism to bring about systemic change, but we also need systemic change to make worker activism more possible and effective. Both the union and non-union workers we spoke to articulated this. They recognize that to be more proactive and bolder, unions need support from other workers and community members, and vice versa. They understand that many of the forces that value our labor differently are the same forces that value our lives differently, and that we need strong coalitions to address the climate crisis, especially in the context of eroding democratic institutions.

We heard over and over from young workers that while climate change is terrifying and overwhelming, collective action can act as an antidote to feelings of anxiety and dread. A teacher in

Massachusetts shared, "I've worked in non-unionized contexts and then I've worked in a unionized context, and I think the main thing is, I just don't feel alone. I truly have an understanding that I'm part of some other larger thing, and that is an important thing to keep front of mind." They added that participation in their union had helped them understand that "there is a possibility to change things, to organize for change, to actually build out democracy."

The Young Worker Network

We started the Young Worker Listening Project because we had a hunch that young workers have huge potential to bridge the labor and climate movements—and to push both movements to be more proactive in protecting workers and communities against increasing crises. Over the last three years, this hunch has become a conviction. We've seen that many young workers are eager to take on this challenging and necessary work, but finding the space to do so is not easy. Young workers face a range of challenges, including lack of capacity and burnout, economic precarity and transience, and a dearth of pathways for leadership development within organized labor. Perhaps the most pervasive challenge to organizing young workers is their uncertainty about the future. In every interview, we asked, "How do you ideally imagine your life and work 20 years from now?" This was the question that young workers most struggled to answer. Many don't know what their employment will look like in six months let alone twenty years, making it difficult to invest in creating the structures needed to organize their communities toward a just transition.

During the interviews, we heard young workers express a desire for dedicated spaces to connect with other young workers who are interested in bridging the labor and climate movements. This demand led to the formation of LNS's Young Worker Organizing

Committee in 2021. In September 2022, the Young Worker Organizing Committee and LNS staff brought together nearly one hundred young workers from across the country to the first Young Worker Convergence on Climate in Los Angeles. At the convergence, young workers participated in workshops and panels on how they're building labor power in service of climate justice. Participants expressed a need for continued connection with a network of young workers in order to strategize across states and sectors and to build a collective vision for a labor-climate movement powerful enough to address both the climate crisis and the inequality crisis.

Through the Young Worker Network, we're creating a shared space of solidarity and learning. By organizing a network of young workers who are eager to act on climate justice, we are attempting to provide them with an organizing structure and, crucially, a sense of belonging and mutual support so that they can build a powerful climate justice–oriented labor movement from below.

We're bringing young workers together around both sector- and place-based issues. For example, members of the International Brotherhood of Electrical Workers union in the Young Worker Network from Los Angeles and Nashville are developing strategies to electrify marginalized communities while creating high-road jobs for members of those communities. We're also bringing together American Federation of Teachers and National Education Association members with transit union workers and environmental justice organizers to develop an organizing plan to achieve resilient, green communities throughout the Midwest by organizing for expanded public transit and transforming school buildings to serve as community resilience hubs.

To ensure a just transition from fossil fuels, we need to put forward a vision for what a better future looks like. Absent that vision—and the organizing power to bring it to fruition—we can expect the transition to resemble deindustrialization: designed to extract as much profit as possible while treating workers, communities, and the environment as disposable. As an electrical worker in Washington State told us, "We talk about a just transition; in the union, a lot of people consider that kind of a joke. It's almost like a dirty word because we know that's not going to happen. I believe that we should never mine another hunk of coal in this country, and I believe that not a single coal miner should be poorer for it. And one of those things is going to happen." That's why the Young Worker Listening Project has invested in developing a powerful narrative by and for young workers—one that sees them as whole workers, that centers care and healing, and that dismantles false distinctions among workers, environmentalists, and community members. In addition, any narrative we put forward about climate justice must integrate environmental, economic, and racial injustice, which a majority of the young workers we interviewed saw as intricately intertwined.

The labor movement is the best chance we have to organize workers, and it needs young workers to organize on the scale necessary to ensure a just transition from fossil fuels. One place to start is by creating the space to listen to and strategize with young workers.

6

ORGANIZING CLIMATE JOBS RHODE ISLAND

Patrick Crowley

This is a story that the media doesn't want you to know about. For decades, their coverage of struggles for worker and environmental justice has painted the two causes as incompatible. This was not always the case.

When the labor and environmental movements were at the zenith of our power in the 1960s and early 1970s, we found ways to link the struggle for worker safety on the factory floor with the fight against corporate pollution. For example, when the U.S. Congress debated the Wilderness Act in 1960, the president of a timber workers' union, the now defunct International Woodworkers Union, testified in favor of the legislation—even though the bill might cost his members jobs in the short term. He knew that if forests were left in the control of timber companies, they would disappear, along with all the jobs.[1] In 1973, when the Oil, Chemical and Atomic Workers (OCAW) went on strike against oil conglomerate Shell, eleven major national environmental organizations urged their members to honor a virtual picket line by cutting up their Shell gas credit cards.[2]

But as both movements retreated in the face of sustained opposition and repression, our fights became narrower.[3] It doesn't have to be this way. The story I am about to tell you is about how the labor and environmental movements in Rhode Island embraced the urgency of the moment and created a new model to organize for a just transition to a carbon-free economy.

Shortly after the presidential election of Joe Biden in November 2020, a group of labor and environmental organizers in Rhode Island started talking about how best to take advantage of the opportunities afforded by the new administration. The organizers agreed that the two most exciting things Biden had talked about on the campaign trail were taking significant action on climate change and governing as the most pro-union president in American history. What was born out of these initial conversations was Climate Jobs Rhode Island, a coalition of nearly three dozen unions and environmental organizations working across the state toward a just transition to a carbon-free economy. We were determined to be ambitious, and, in little more than a year, we helped pass landmark legislation, including laws requiring Rhode Island to purchase 100 percent of its energy from renewable sources by 2033, mandating that all renewable energy projects be built using labor peace agreements and pay prevailing wages, and putting Rhode Island on a path to be carbon neutral by 2050. What we learned is that by thinking big and identifying shared goals, it is possible to unite the labor and environmental movements around a progressive, pro-worker agenda.

Labor unions and environmentalists have long clashed in Rhode Island, as they have in many places around the country. In 2019, for example, the environmental movement successfully blocked the development of a new energy-producing facility in the rural town of Burrillville, Rhode Island, because it relied on fossil fuel technology. Their opposition antagonized the unions of the Rhode Island Building and Construction Trades Council, a key affiliate of the Rhode Island AFL-CIO. The trade unions, still trying to recover from the job losses of the recession following the 2008 financial crash, were incensed. The project promised not only hundreds of construction jobs but also dozens of union production jobs once the facility was complete. From labor's point of view, the environmental movement was running an anti-union and anti-worker campaign. From the environmentalists' perspective, the labor movement was standing in the way of progress toward decarbonizing the state's economy. It was clear we had a lot of work to do if we were going to build an effective labor-environmental coalition.

One of the most important lessons I've learned in my two-plus decades in the labor movement is that, when building a coalition, it's crucial to look for something that can unite the coalition partners. In a workplace organizing drive, this is usually shared antipathy toward the boss; in a community organizing drive, it might be shared values like fairness or respect. In this case, the answer lay four miles off the coast of Rhode Island, in Block Island Sound, the home of the nation's first offshore wind farm. The Rhode Island Building and Construction Trades Council was central to getting the five-turbine, 30-megawatt project approved and completed. The experience of the Trades Council with the offshore wind industry seemed like the perfect way to overcome the antagonism between the two movements. It demonstrated that the labor movement was not inherently opposed to working on clean energy projects, and it gave the environmental community a chance to demonstrate that the renewable energy sector could be a place to build and sustain long-term employment for union members.

To help define our project, the steering committee agreed that we should work together to pass a bill called the Act on Climate, with the goal of creating a 100 percent net-zero emission economy in Rhode Island by the year 2050. We also agreed that if the Act on Climate was to be successful, it needed to call for a just transition to a carbon-free economy. That meant not only that the jobs of the future should be good-paying union jobs but also that those work opportunities should be opened up to the people most affected by pollution and climate change: low-income communities and, especially, Black, Indigenous, and people of color (BIPOC) communities.

Priscilla De La Cruz, a well-respected environmental organizer from Providence, Rhode Island, was especially critical in helping union members understand the impacts of fossil fuel pollution on communities of color. Neighborhoods with majority minority populations in areas around industrial activity, particularly near the Port of Providence, experience higher rates of childhood asthma. Communities in parts of the capital city, meanwhile, have little tree cover, leading to higher summer temperatures and more heat-related illnesses. Priscilla helped all of us in the labor movement see that our core beliefs about worker and economic justice were also racial justice issues.

Michael Sabitoni, president of the Rhode Island Building and Construction Trades Council and a leading labor voice in the early organizing committee work, helped the group understand the importance of ensuring that the jobs in the renewable economy be good union jobs. Generally, jobs in renewable energy are non-union and lower paid, while jobs in the existing economy are highly paid union jobs with great benefits. Compounding the problem is the fact that the current construction workforce is predominantly white, male, and aging, while the emerging workforce is more diverse.[4] If we truly want to build a just transition to a carbon-free economy, Michael argued, we need to make sure the new-economy jobs are unionized so that employers cannot exploit the emerging workforce just because it is made up of more Black, Brown, and female workers.

The real organizing work began as we hashed out these kinds of principles and goals. This foundational work was challenging because the issues raised were often brand-new concepts for people in both camps. We didn't even understand the language each other used. For example, our friends in the environmental movement had heard of project labor agreements, but they didn't really know what they were. Folks in the labor movement, meanwhile, were familiar with words like *resiliency* and *carbon neutral,* but we didn't really understand how the environmental movement used these terms.

To get the coalition groups on the same page, the steering committee created a series of shared learning experiences. The first, a breakfast meeting we called "Davis Bacon and Eggs" was held in early January 2021. As the labor groups in the coalition knew, the Davis-Bacon Act is a 1931 law that requires that all work performed on federally funded public works projects be paid the prevailing wage for the local area. In Rhode Island's case (the state also has a version of the Davis-Bacon Act), this means the union wage must be paid for the various construction trades. Our second shared learning experience focused on the transportation industry, since nearly 40 percent of Rhode Island's fossil fuel emissions are generated by transportation. The idea of making the Rhode Island Public Transit Authority (RIPTA) fare-free and powered by 100 percent clean energy buses emerged from this learning experience. Since the workers at RIPTA are members of several different unions, primarily the Amalgamated Transit Union Local 618, working to expand RIPTA services would not only take more cars off the road, it would also lead to an increase in union membership.

These shared learning experiences helped us to develop new sets of vocabulary while building trust among groups that had no history of working together. Additionally, they gave more people an opportunity to assume leadership responsibilities and showed that the steering committee was intent on building a movement. As both an organizer and a participant in these sessions, it was amazing to watch so many "aha" moments and see how they drove us to create a common agenda.

In late January 2021, Climate Jobs Rhode Island officially launched. We published a website, www.climatejobsri.org, and held a virtual press conference broadcast over social media channels. The event was well received by members of the state legislature and garnered a little notice in the press, but, more importantly, it got the attention of the Climate Jobs National Resource Center (CJNRC). Within days of our press conference, the president of CJNRC, Michael Fishman, the former secretary-treasurer of the Service Employees International Union, called me at my office in Providence to say he had heard about our work and was interested in collaborating. According to Michael, Rhode Island was on CJNRC's target list of states for a possible state-based coalition. Since we were already well ahead of them, he asked, how would we like to have some financial support for our work?

The answer to that question is always yes. So, with the help of CJNRC, we began to put our plans into action. The first step was to pass the Act on Climate to establish the overall goal of a just transition and then to build the legislative scaffolding to make it meaningful and enforceable. We worked with Senator Dawn Euer, the chairwoman of the Senate Committee on Environment and Agriculture, and Representative David Bennett, the chairman of the House Committee on Environment and Natural Resources and a retired union nurse, to modify the Act on Climate to include language addressing workers' issues. We also turned out to support the bill at legislative hearings. It was exciting—and unprecedented—to see statewide leaders of the labor movement standing shoulder to shoulder with leading environmentalists. Legislators who were used to seeing labor and environmental activists testify against each other now saw us sharing the same talking points at bill hearings and speaking together at press conferences and rallies. Over an intense couple of months, as members of Climate Jobs Rhode Island lobbied continuously alongside one another, the Act on Climate finally worked its way through the legislature. Rhode Island governor Daniel McKee signed the bill into law on April 10, 2021.

Despite our early successes, none of this was easy. In addition to learning how to have two very different movements work together, we had to overcome obstacles thrown at us from other groups, including competing advocacy groups. For example, even though the Sunrise Movement in Rhode Island had been a supporter of the Act on Climate when it was first introduced in 2020, by 2021, they voiced opposition to it, claiming it didn't go far enough. This stance really came out of nowhere: the language of the bill had only been strengthened over time. In my opinion, Sunrise's newfound opposition stemmed from its alignment with a coalition of purportedly progressive political candidates called the Rhode Island Political Cooperative, known simply as the Co-op. The Co-op had been formed to support candidates challenging elected leaders in the state legislature from the left. Despite Rhode Island's reputation for being a liberal state, many conservatives run for elected office as Democrats. While that was less true in 2021 than it was a decade earlier, it was still the case that the Democratic Party's "big tent" sometimes let the elephants in. However, for reasons that baffled many political observers, the Co-op didn't threaten the more conservative members of the legislature with primary elections; instead, they started to put forward candidates to challenge avowedly pro-union and pro-environmental legislators, including several key supporters of the Act on Climate. Worse than that, some of the initial candidates they promoted held deeply disturbing views, including rejecting reproductive freedom and supporting vaccine conspiracy theories. The Co-op's work made the politics of moving our agenda forward that much harder. If supporting legislation benefiting working people and the environment didn't insulate a lawmaker from a left-wing challenge, how could we convince the more conservative members of the legislature to support our bills?

The antidote was to keep our coalition strong and focused on the same goal, which we were able to do because we had established trust and had collectively developed shared principles early on. The Co-op challenge fizzled in the end, and in the next legislative year, we were able to add to our early success. In January 2022, thanks to CJNRC, the Industrial and Labor Relations Worker Institute at Cornell University produced a report for Climate Jobs Rhode Island outlining how the Ocean State could meaningfully pursue the goals of the Act on Climate.[5] Following the publication of the report, we worked to pass legislation that required Rhode Island's utility to purchase 100 percent of its energy from renewable sources by 2033; required the state to license up to 1,000 megawatts of new offshore wind projects; mandated prevailing wage, apprenticeship utilization, and labor peace accords on any renewable energy project over 3 megawatts; and established a fare-free pilot project on the RIPTA bus system's most utilized service line. Additionally, the speaker of the House of Representatives appointed me and Priscilla De La Cruz to serve on the advisory board monitoring the implementation of the Act on Climate.

Five years ago, if you told me that the labor movement and the environmental movement would be working together to establish a just transition to a carbon-free, worker-friendly economy in Rhode Island—that we would work together to pass laws establishing the nation's most ambitious 100 percent renewable energy standard and take a big step toward providing free public transportation throughout the entire state—I wouldn't have believed you. But that is exactly what we did. We started out looking to be ambitious with our goals, and our approach has paid off. We didn't have a model for how to do this when we started, but now that we do, it is our hope that organizers in other parts of the country will adapt and strengthen our model so that we can create a pro-worker, pro-climate economy everywhere.

Part II

KILLING THE WIINDIGO

CLIMATE JUSTICE, WORKER JUSTICE

7

HOW TO WIN A JUST TRANSITION

AN INTERVIEW WITH JOSE BRAVO

Jeff Ordower

Jose Bravo is the executive director of the Just Transition Alliance (JTA), a coalition of environmental justice [illegible] and labor unions that was founded in 1997.

What is the origin of the phrase *just transition*?

About twenty-seven years ago, when we first talked about just transition with the Oil, Chemical and Atomic Workers Union (OCAW), the idea was to transition workers from unsustainable production to something that's sustainable and actually better for health, the planet, and the work space. It was really, really worker centered. The environmental justice movement was brought into the space after the workers initially thought about bringing in mainstream greens. Then as an afterthought, basically, they said, "Well, it didn't work out with the mainstream greens, but we hear that there's this environmental justice movement that we might be more natural allies with." The first meeting was among the OCAW and the Southern Organizing Committee, the Southwest Network, the Indigenous Environmental Network, the Asian

Pacific Environmental Network, and Farmworker Justice, so a good swath of the environmental justice movement.

What were the early relationships between unions and environmental justice groups like?

Initially, there were a lot of challenges with unions because there was a lot of white worker privilege at the table. We were told that if the OCAW saw us picketing outside or trying to shut down a plant, that wouldn't be permissible. So environmental justice groups said, "Well, then you have to take away your right to strike." And they said, "Well, no." Our response back was, "If you give up your right to strike, then we'll give up our right to have direct action." So we developed a mutual agreement to inform each other about actions.

The other thing that workers were saying was, "We see our jobs in jeopardy, and there are all these companies that do contracting and some of these contractors move in on our jobs." So we said, "Wait, let's unpack that." Because we knew that temporary agencies hire these contract workers, who usually don't speak English. They hire folks who don't speak English, aren't trained, and then they do the jobs that union members usually refuse to do: cleaning those tanks, putting themselves in danger without really knowing it. Union workers can refuse to do certain work and we can't. So we said, "It is in your best interest to actually hire some of these folks or unionize some of these folks."

So what happens to a contract worker who doesn't speak English, who gets to do some cleanup and doesn't even know what they're handling? Well, they get sick. They don't have any other recourse because they work for a temporary agency. There's no insurance.

So what would be a just transition? Frontline communities wanted to transition into good livelihoods. And we were thinking about who's going to clean up the mess, the legacy of contamination from the fossil fuel industry and what it's caused in communities of color, low-income communities, and Indigenous communities here in the United States. OCAW meanwhile had been dealing with lockouts and collective bargaining that had been at a stalemate for two or three years.

We decided to focus on several places where we could develop worker-community campaigns: Ponca City, Oklahoma; Rillito, Arizona; San Antonio, Texas; and Macintosh, Alabama. We started to see that there was one common denominator that joined us all, and that was worker health and safety and community health and safety.

At the beginning, it was about working on trust. To be able to say, "Hey, I trust that you'll have my back. And yeah, I'll be out at that line working for you on your contract issue as well. So I have your back."

Doctor Marvin Legator with the University of Houston put together this survey with 336 questions. We had community members survey workers; we had workers survey community members. And it turned out that some of these community members knew these workers. They went to the same high school, they played on the same football team or did different things together. It was just that they had lost touch. So that put us into a situation where workers and communities came together. The other thing that we pushed from the community side was that worker health and safety meetings no longer be held inside the plant—that they be in the community.

And that absolutely drove management apeshit crazy because, at that point, workers were talking about health implications in the community instead of safety inside the plant. Workers on management's side would go to these meetings, and the other workers would say, "Get the hell out of here. We know who you are. You're not a union guy. You pay your dues, but you're being a spy for management."

In the end, we were able to settle every single one of the worker demands that we focused on, and we were also able to get the Arizona Department of Environment to address the issue of off-site contamination from the plant itself.

What has your relationship with labor been like in recent years?

Unions were getting attacked from all sides, especially with right-to-work legislation. The OCAW was a left-of-center union, and it was a radical union compared to many other unions. And they were literally the ones that came to us and said, "We see the writing on the wall. Things that we make probably don't belong on the face of the planet, but how do we transition into something new?" But OCAW was pretty small, so they merged with the Paperworkers to become PACE, the Paper Allied Chemical and Energy Workers. And the PACE side of things was much more conservative.

The first meeting that we had with the PACE leadership, the president and the vice president looked at Tom Goldtooth, head of Indigenous Environmental Network, and they said, "We are not going to have the discussion around dioxin. If you want to have the discussion around dioxin, this is over. It's gone. Let's drill down on worker health and safety. Let's drill down on trainings, other things like that. Let's build this." So that was the reception that we got from PACE. And that went on for several years, and then all of a sudden, here we go with another merger, and we started to see that there was interest from the United Steelworkers to merge with PACE. As the labor movement consolidated more, we began to work much more deliberately with local unions so as to build place-based relationships.

What locals have you had success working with?
My favorite campaign was the Rillito, Arizona, campaign. In Spanish, *rillito* means "little river," but there is no river there; there's a creek. There's probably about fifty or sixty households in Rillito and then a giant looming plant with giant kilns that you can see for miles. Before you see the community, you see the smokestacks going off. That community of Rillito, Arizona, had probably the largest concentration of African American people within a community that I had seen in Arizona. And the community was just like any other environmental justice community: it was devastated by that big-ass plant right next to the fenceline.

Every Wednesday, the workers would do a picket there because their collective bargaining contract hadn't been signed. And the community, those people that live in those fifty households or such, would just drive by. And David Garcia, the president of the local, never really thought about just going to Rillito because their issue was their issue and the community's issues were their issues. So when we came, we brought the workers and community together in the community center. And then people would say, "Hey, I remember, I played football against you. Hey, we played baseball."

When we started bringing folks together, the first thing that we had them do was a community cleanup. The workers had access to tractors and big giant skip loaders and things like that. So they said, "Yeah, we'll bring them out. We'll come out and do that." So that in itself was a move in the right direction because it literally brought some trust to folks.

And then we told the community, "Well, how come you don't stop by on Wednesdays and support your friend that you played baseball with?" And so they started coming together and helping each other out. Then the workers, when we started talking about the health and safety issues, the workers came to us and said, "Hey, we know where the bad stuff is buried because we buried it. All the chemical drains, we were asked to bury them, and they're buried right there." And they brought air pictures taken from an airplane and stuff, and they said, "See, here's the community, here's where we buried all that stuff, and we believe that it's affecting us."

The locals in LA were also supporting us. The central offices for Arizona Portland Cement were in Altadena, California, so we would bring workers and community members to do actions outside the central offices. And likewise, they would make puppets for us, giant heads of Dave Bittel, who was the plant manager, and float him outside the corporate offices, that kind of stuff.

And then something happened inside the plant. There was what's called an *upset*. I don't know if it was an explosion or a leak or something of that nature. The state of Arizona moved in and sanctioned the plant $600,000. So we said, "Okay, state of Arizona, now you have the money. We don't want you to put it into the treasury. We want you to develop a site environmental project along with the EPA." So the EPA came in and we put monitors in the community with that money. And we had the community and the workers trained to read the monitors and read the off-site consequences from that plant. So every day, folks would actually take all those records and then make sure that they were tallied and other things like that. They were trained to do this. And then at the end of the month, all that information was sent back to the state and the federal government. That put a lot of pressure on the company. The company also agreed to lower their conveyor belts and put a water system on their conveyor belts because the dust was really, really bad. And ultimately, the workers got their contract, the community got their cleanup, and the community got access to a health clinic. It was just good work.

How do you do just transition work at scale?

We went to Sharm el-Sheikh for the COP negotiations this year. It was really good to see the work that has happened for many years where the folks who actually push the buttons and turn the knobs are talking about false solutions and are talking about making sure that we don't go down the path of worsening conditions both in communities and in workplaces.

Just transition is not a cookie-cutter approach. It's not one thing for everyone. But I can tell you beyond a shadow of a doubt that if a just transition doesn't have workers and there's only communities at the table, then it's not a just transition and vice versa. If it only has workers and the community's not at the table, then it's not a just transition. A just transition is literally a cradle-to-grave approach that removes the exploitation out of the whole process of production.

We need to take a just transition to scale. Right now in the United States—here we are almost in December—this administration is pushing us toward electric vehicles, electric vehicles, electric vehicles. In order to meet even the simple goals of the Paris Agreement, and I call them simple because I believe it should be more around 80 to 85 percent reduction of carbon emissions instead of whatever was agreed to, we need to do much more than manufacture electric vehicles. A step in the right direction in a just transition would be to scale up mass transit systems. The subsidies that the oil and chemical companies get should be redirected so that mass transit can be free. We would like to see Detroit retooled into making mass transit vehicles that don't pollute and become the world hub of that manufacturing. That's a very clear example of how the United States needs to move forward.

The Biden administration is touting, "Oh, we have 500,000 new plugins for EVs." In my community, there's not going to be that many EVs. They're too damn expensive. It's not going to work for certain communities that have historically been left out. But I'll tell you what, if you make mass transit free, guess who gets to use that infrastructure? So develop the infrastructure for adequate mass transit. Move us toward reducing carbon emissions instead of just paving the road with false solutions.

8

ORGANIZING COAL COUNTRY

Veronica Coptis

When I was a teenager, a coal mine's waste dump site expanded next to my family's home. I spent years of my life angry, wondering who would speak for the ecosystem that was now filled with toxic waste. *The coal company gets what it wants*, everyone told me, *and some people in our town get jobs out of it*.

I went to college to study biology, with a focus on wildlife habitat. But after finishing my degree during the Great Recession, I found myself back at home, waiting tables at the local diner. That experience opened my eyes to the power of the coal industry in my small town in Southwestern Pennsylvania. I served the people whose homes were destroyed by longwall mining as well as the men and women who depended on those jobs to support their families. I realized that my activism against the coal industry was born out of privilege. No one in my family depended on coal to make a living.

I knew there had to be a way to fight for the needs of the workers and the people living above the coal mines, but I had no clue where to start. Then a coal mining and fracking activity caused a forty-three-mile fish kill in a nearby stream. A regular at the diner asked me for help getting accountability on the fish kill, and together, we found the website for the Center for Coalfield Justice

(CCJ), an organization that fights for community members in coal-dependent communities in Pennsylvania. They were hiring for an AmeriCorps position, so I took a shot and applied. Seven years later, I became the organization's executive director, a role I held for the next six years.

CCJ was founded in 1994 as the Tri-State Citizens Mining Network—a coalition of grassroots organizations in Ohio, West Virginia, and Pennsylvania that advocated for state mining programs to protect residents from the impacts of longwall coal mining. At the time, longwall coal mining was a new mining technology that removed the entire coal seam, leaving little support for the ground above and causing significant damage to water supplies, structures, and streams. The new technology not only harmed the environment, it also cost workers jobs because it required a much smaller workforce to operate. In fact, the shift to longwall coal mining is responsible for many of the lost coal mining jobs in the 1980s and 1990s.

The Tri-State Citizens Mining Network took an environment-first approach to organizing rather than attempting to forge an environment-labor coalition. In fact, for most of our thirty-year history, CCJ has valued natural resources far more than good jobs. Since 2014, we have been building organizations *with* workers—rather than in opposition to them. The assumption that fossil fuel industry workers don't care about the climate crisis is a fallacy. These workers understand that our climate is changing and that it's going to affect them, their kids, and their grandkids. Many of these workers are also living around the extraction operations: their water is poisoned by pollution, and they experience the public health impacts of mining. Rather than dismissing these workers, organizers need to engage them by listening deeply to their concerns and by finding common ground.

CCJ's role is unique: we are fighting for the environment and for the workers. It is challenging, but we continue to engage on both fronts. We believe that we can—and must—build alliances with workers who fundamentally disagree with our vision, if we do our research, tell the truth, and, most importantly, meet workers where they are at, with curiosity and an assumption of mutual interest. Everything CCJ does boils down to how we engage people in the organizing, whether they are workers or community members.

Climate is an issue that people care about across political boundaries. All working people have a role to play in organizing to fight climate change, so we must create as big a tent as possible. It has taken us a long time to get to where we are, and we still have a long way to go, but here are some lessons we have learned about what it takes to build a climate-labor movement in coal country.

Lesson #1: Meet people where they are at—and be willing to have hard conversations and lean into conflict.

In 2016, CCJ took the Bailey Coal Mine to court to protect a state park that is critical to our state's economic future beyond coal. A portion of the mine's work would have damaged nearby streams beyond repair, threatening a critical water source that might be essential for new economic work in the future. We filed for—and won—an emergency injunction to stop that tiny chunk of the mine from moving forward. We never called for shutting down the whole operation, but the company began to lay off workers and held press conferences blaming CCJ for the layoffs. Workers then called our organization, fuming, "What am I going to do? This is my job. How do I feed my family?" In response, we put up a Facebook post inviting workers to call us to learn more about the permits, as long as they didn't verbally abuse our staff. When some of them called, we validated their concerns and then told them that the company still had a permit to mine the next two panels—meaning they would likely return to work in the coming weeks. Being a repository of accurate information helped us to earn a lot of trust with the workers. In that moment, the workers learned that CCJ operated in good faith and the boss did not.

Because of the diminishment of unions, many workers do not have a basic mistrust of the boss, meaning we have to tread more lightly. We do not enter conversations with workers saying, "The boss is horrible." Instead, we have initial conversations to assess where each worker is at. Our organizing rap looks something like this: *What do you like about your work? What don't you like about it? You're talking to me so you must care about the environment. Tell me what you're concerned about.*

Organizing in this way means we have to train ourselves in how to navigate conflict. If we are actually going to start building a multi-issue, multiracial, multiclass movement, there will be a ton of conflict and misalignment along the way. For example, we recently met with one of our members from a more affluent, conservative-leaning community about joining our board. During the conversation, the member revealed that, while they believe in climate change, they do not think that it is caused by human activity. CCJ exists to fight human-caused climate change, and so we told the member that they couldn't join the board if they were not aligned on that mission. But we didn't leave it at that. We asked questions about their beliefs and shared information about our own. In the end, we told them that we would continue to support them in their fight to stop a mine from being developed in their community. One of the keys to navigating conflict while continuing to build power is to not abandon people when there is disagreement.

Lesson #2: Listen to workers—even if they don't agree with you—and invest in building up their organizing capacity.

Building a big tent starts with listening to workers, even when they don't agree with us. CCJ didn't begin to think about the impact of mining practices on workers until 2013, when CONSOL Energy Inc. sold some of its mines to Murray Energy. The company was going to cut health care benefits for retired miners even as it promoted itself as a caring member of the community. We published a letter to the editor of the *Observer Reporter* in which we called out the corporation for cutting benefits.

After the op-ed was published, some of the mine managers reached out to us for a meeting. They were skeptical of CCJ, which they thought cared only about water, animals, and climate—not workers. They didn't agree with us on the need to combat climate change or to regulate the coal industry. But we both agreed that mine managers should have their health care covered through retirement, like the company promised. So we worked together to co-create a campaign to force CONSOL to reinstate the mine managers' health care benefits.

The key to that campaign was that the mine managers made every decision in conversation with CCJ. We taught them how to analyze the situation: how to identify decision-makers, how to escalate actions to get the target to respond. But they decided which actions to take. Our investment in building up their organizing capacity paid off. When the CEO and head of human resources for CONSOL Energy met with the miners, they warned the miners against working with CCJ. But the miners pushed back, telling them that we were the only group—including unions—that had met with them and given them the tools to organize themselves. What we learned through that campaign was that we needed to listen to our communities, including the people who don't agree with us.

After the CONSOL campaign, we decided to ramp up our economic justice work. In 2019, we ran a deep canvassing campaign with the goal of knocking on over one thousand doors of households in Greene County with an income of less than $60,000. We didn't want to talk only to those who were ideologically aligned with us but to reach all of our neighbors who were financially insecure. Through the canvass, we learned that a majority of residents in Greene County are not working in the fossil fuel industry. Most residents are concerned about energy issues but even more so about how increased cost would impact them. They overwhelmingly believed that coal supported our local tax base, but they did not believe the same of the gas industry. That canvass gave us a better understanding of where our communities stood on a variety of issues, informing our future campaign work and analysis.

Lesson #3: Talk to unorganized workers—not just union members.

When we talk about workers, we too often focus on organized labor. The reality is that most of the fossil fuel sector is unorganized. Because these workers have no protections, they are much more fearful about speaking out. Most of the workers who reach out to CCJ are not part of a union but work or live near active fossil fuel operations. These workers understand that companies can do better to respect and protect communities, but they also know that if they speak out publicly, they can get fired. But we still need to find a way to center the experiences of these workers and their families.

One way we seek to incorporate these workers' voices—without risking their employment—is through our economic justice listening sessions. In 2019, we invited everyone we canvassed to talk collectively about how to improve our local economy. Over sixty people attended the listening sessions, two-thirds of them non-union. Notably, most of them agreed that we need to plan for a shift in the coal market, that unions should play a key role in organizing, and that we should protect pensions and health care. Supporting the unionization of the remaining fossil fuel workers could be a critical organizing drive to build ties between labor and community.

Another way we seek to incorporate the voices of non-unionized workers and community members is to hire staff exclusively from the communities we're working in. That means that our staff live on the front lines of the extractive economy, and we have friends, family, or loved ones who work in these industries. My husband, for instance, worked as a coal miner when I started at CCJ. These trusted relationships allow us to have confidential conversations with fossil fuel workers to learn what's happening on the ground—and to use that to inform our strategy.

Lesson #4: Center political education.

At one of our economic listening sessions, we asked the question, "Do you believe the minimum wage should be increased to $15 per hour?" We then had workers stand in different areas based on their level of agreement or disagreement. A lot of our younger members who were working in the service sector expected older white miners to oppose raising the minimum wage. But the United Mine Workers stood with our young members, while non-union, middle-class members opposed raising the minimum wage.

When we asked participants to explain why they were standing where they were, the United Mine Workers talked about the importance of being in a union. The political education that workers get in unions goes beyond workplace issues—it shapes how they show up in their communities. The decline of unions, meanwhile, has meant that many workers get their information from whatever media outlet they have access to.

After that listening session, we decided that political education had to be a key component of our work. We started hosting monthly community meetings, many of which have popular education style learning built into them. We pick an issue each month and invite our members, supporters, and anyone from the community to discuss it together so they can learn from everyone's experiences. Through these sessions, we have seen members' analysis deepen and their ability to advocate for themselves grow.

Lesson #5: Organized labor needs to be bolder, but environmental justice groups also need to stand with labor.

Many labor leaders in coal country are just holding on to what they have. They want to make sure their members maintain their jobs and get the best benefits possible. But most aren't even trying to grow their membership, let alone fighting for broader political or environmental change. Too often, I hear these leaders complain about how labor law has been gutted or contractors have flooded the market. But this region has a rich history of labor fights—and none of them have been easy to win. The question is: What are we willing to risk for the rights of workers going forward?

On the other side, environmental justice groups need to stand with labor. Workers have never won alone: communities, families, and churches must show up to back them. Too many climate groups ignore workers in their quest to reduce carbon emissions. For example, community solar is illegal in Pennsylvania. In 2018, the state legislature introduced a bill to make community solar legal. The bill was being pushed by solar companies in conjunction with environmental groups. But there was no requirement that solar installers be unionized. When CCJ raised this issue with the Clean Power PA Coalition, the state and national environmental groups in the coalition refused to stand with the workers, because they'd lose the support of the solar companies backing the deal. In the end, the bill did not pass, and environmental groups missed a critical opportunity to build worker and community power.

Conclusion

Too often, we talk in platitudes about how workers and community members should just sit down and talk. But talking is insufficient. What we learned at CCJ is that standing in solidarity with workers means we have to listen deeply and ask questions, even when we strongly disagree with what they are saying. We have to face workers' anger and fear. Most of all, we have to be prepared so that we can act as trustworthy sources of information when the boss lies, as the boss always will. This work is slow. It takes time and patience. There will be many missteps along the way. But we have no option but to forge a path that brings workers and frontline communities together. If we don't, we will be left out of the policy solutions pushed by the mainstream environmental movement. To shift away from an economy that has extracted our resources, labor, and culture, we must begin to see ourselves as a greater "we." Together, we have the power to develop a new future, with thriving communities that meet all of our needs.

9

"RESILIENT COMMUNITIES ARE ORGANIZED COMMUNITIES"

AN INTERVIEW WITH VIVIAN YI HUANG AND AMEE RAVAL

Miya Yoshitani

Working-class communities of color are experiencing the cumulative impacts of converging public health, racial, economic, and climate crises. As states like California face devastating wildfires, extreme heat, power outages, and an ongoing pandemic, the need to proactively advance climate adaptation and resilience is clearer than ever. However, climate resilience efforts typically focus on improving hard infrastructure—such as roads and bridges—to the detriment of social infrastructure: the people, services, and facilities that secure the economic, physical, cultural, and social well-being of the community. Traditional models of disaster planning have also proved deeply inadequate. They are coordinated through militarized entities like local sheriff's departments and rely on protocols like evacuating to faraway and unfamiliar sites, sharing emergency alerts in only one or two languages, and requiring people to present identification to access services, thus shutting out many from the support they need.

Through these crises, new models of disaster response have emerged. In some places, neighbors have formed mutual aid networks to share resources with one another; in others, schools have provided food to tens of thousands of families. During extreme heat waves, libraries have been turned into cooling centers. What these approaches have in common is that they are rooted in the existing social and public infrastructure of communities. They are, in effect, resilience hubs.

Resilience hubs are physical spaces where residents and community members can access resilience-building social services on a daily basis as well as gather, organize, and access response and recovery services during disasters. The creation of resilience hubs can also provide high-road job opportunities in infrastructure building and service delivery.

The Asian Pacific Environmental Network (APEN)—an environmental justice organization with deep roots in California's Asian immigrant and refugee communities—has been advocating for policies to scale community-driven resilience hubs across California. Vivian Yi Huang is the co-director of APEN and Amee Raval is the policy and research director.

Tell me about the history of the Asian Pacific Environmental Network.

Vivian Yi Huang: In 1991, there was a National People of Color Environmental Leadership Summit, a seminal event in the history of the movement for environmental justice. There were a number of Asian American organizers and activists who really recognized the need for an organization to focus on environmental justice in Asian communities. APEN grew out of these conversations.

We center the leadership of frontline community members, and we also put forward the idea that the environment does not live separate from us. It is really about where we live, where we work, where we play, where we thrive. And that was a huge shift we wanted to make in the fight for environmental justice. Working class Asian immigrant refugees are on the front lines of pollution, toxins, economic inequality, gentrification, and displacement. Over time, we started to recognize that the policy decisions at the state level had an impact on whether our members had access to clean air, whether our members had access to health care and housing, whether our members were actually going to see investments in their communities, so we began to do state-level climate policy work as well.

APEN has grown from a local, community-based environmental justice organization to having some critical policy wins at the state level. Can you share some of the climate resiliency policy wins from the last few years?

Amee Raval: APEN has spent decades fighting for a local just transition away from polluting industries toward a regenerative economy that can support the health and well-being of our community members. A few years ago, we started looking for ways to intervene at a state level through the lens of "climate resilience" to make that happen.

One of APEN's earlier efforts into climate resilience policy was about improving the state's capacity to identify climate vulnerability at a community level. The state of California has multiple agencies mapping and isolating different climate impacts, like sea level rise, wildfire risks, or extreme heat. But there was no comprehensive tool to identify the cumulative impacts of these different hazards to locate the neighborhoods that will be hardest hit by climate change, like the environmental justice mapping tool CalEnviroScreen does for environmental justice issues. APEN researched and wrote a *Mapping Resilience Report* that helped to look at where the hotspots of climate vulnerability are, not just from the standpoint of climate disasters like flooding, heat waves, and drought but overlaid with socioeconomic inequalities and community conditions that strongly indicate whether or not communities are able to respond in times of climate disaster—and if they have the infrastructure and access to the resources they need to prevent them from happening in the first place.

We also started to bring together different stakeholders who would benefit from building the new climate resilient infrastructure that communities need to survive and thrive through climate crises. Out of that, APEN started working in coalition to advocate for state resources for community resilience hubs. These are community-owned buildings and spaces where people can access response and recovery services during disasters, they can get important social services on a daily basis, and they can use the space to gather and organize.

Through the past couple of years of advocacy, we were able to secure $270 million to fund community resilience hubs. We also helped secure over a billion dollars for investments in healthy homes, which ties weatherization and other home-resilience upgrades with housing affordability and anti-displacement measures. We were able to get a federal earmark of $2 million for a local resilience hub project in Oakland's Chinatown, at the Lincoln Recreation Center.

And finally, I want to uplift our community solar work through a bill we worked on last year, AB 2316, which is a community solar program in California that's designed for working-class communities. The Bay Area residents we work with, most of whom are renters, can actually see projects in our neighborhoods on a library, a temple, or a parking lot—cooperative, community-owned models for solar that create jobs and increase local ownership of clean energy infrastructure.

So climate resilience is not just about getting rid of pollution, displacement, and gentrification—the inequality that people

face on a day-to-day level. It's also about what people really want to see if their communities were invested in.

Raval: What do our communities need to weather the storms, not just climate but all of the crises—like gentrification, growing economic inequality, the pandemic—that we're facing at once? We can't think of these in siloed ways because they're happening all together. We had our leaders design the Healthy Homes bill, which looks at more than resilience and weatherization investments but at the impacts on renters. Not all investments are positive if they're going to make housing unaffordable to current renters and drive harms like gentrification and displacement. Our climate resilience work has really been an opportunity to create the vision of what we want to see as part of a regenerative economy. What are the new systems that we want to build? There's community ownership and governance of land, housing, and resilience. But our leaders wanted to show concrete changes at the neighborhood level: models for renewable energy and solar and battery storage, investments in programs like the Solar on Multifamily Affordable Housing program [a $1 billion investment over ten years in solar systems built on multifamily affordable housing in environmentally and economically disadvantaged communities in California]. We want to build models that take what is currently the norm and apply the principles of cooperative work, regeneration of resources, deep democratic governance, as well as centering ecological and social well-being.

It sounds like you're saying there's different layers to climate resilience work. One is the actual infrastructure and the investments—the hard infrastructure that calls for billions of dollars of investments in communities that need it the most. But then there's also the investment in the social infrastructure that is part of how our communities become

more inclusive, more democratic, more connected, more prepared and powerful. Sometimes people see communities as inherently resilient and don't think they need as much investment. How have APEN communities responded to the term "resiliency"?

Raval: Immigrant and refugee communities and working-class communities of color have had to build resilience for themselves, because of multiple stressors and inequalities, just to survive. But being labeled "resilient" because we are still here doesn't mean we don't need a lot of investment in our communities. It just means we've had to navigate simultaneous impacts. These are communities that have been historically disinvested from; these are the places that need to be invested in now. We should also get to self-determine the ways we want to talk about the fight for community-led transformation. RYSE [Youth Center] reframed it as a "liberation and resilience hub." That brings in more of a power-building, organizing, dignified lens.

Resilient communities are organized communities. During the pandemic, we saw young people translate important updates for their parents and share resources with their peers. And we saw high school students volunteer to deliver food to elders in their communities. When there were refinery flares, they alerted their friends and families to stay inside. Resilience requires building strong relationships within and across our communities.

It also means challenging these big environmental groups to shift their thinking about climate adaptation and resilience. A lot of adaptation efforts focus on what to do to protect forests, coasts, wetlands, and those big infrastructure projects like sea walls, roads, and bridges. But our communities remain vastly unprepared. We need those deeper investments in the systems that strengthen the social and economic fabric of our communities so that we're better resourced in times of crisis.

Can you tell me more about the concept of a resilience hub?
Huang: Resilience hubs are trusted spaces where people already gather—like a youth center, a school, library, clinic, or a place of worship—and that people turn to for support and help during natural disasters like storms, wildfires, heat waves, and power outages. These places are already part of a community's social fabric, and the idea is to invest in them rather than create entirely new infrastructure. The vulnerabilities of each community are different, so this is not a cookie-cutter approach, more like a menu of options to strengthen the resiliency of that particular neighborhood. For example, one option could be installing a solar and storage microgrid. Another could be installing rainwater storage tanks and filtration. Or, putting in efficient refrigeration units, or air conditioning and proper air filtration systems, electric vehicle charging stations, communications hardware, or physical storage space for distribution of emergency supplies. And most of these structures would also need building upgrades, weatherization, earthquake retrofitting, or other improvements.

As Amee said before, "Resilient communities are organized communities," and these hubs are also potential centers for "social infrastructure" for social resilience and better cohesion of a community. As converging economic, political, and climate disasters become more frequent and intense, we need to build power and develop collective responses to disasters, so that our communities can weather the storms together. APEN has imagined community resilience hubs could add capacity for activities like voter registration, workforce training and development, youth services, health services, legal services, and housing support. They could also provide space for community meetings and organizing. And these hubs can be models for local ownership and governance that give our communities more agency and decision-making over the design and the priorities of the resilience hubs.

For example, in Richmond, disasters like refinery fires, oil spills, and power shutoffs are a constant threat. Our communities face decades of disinvestment from schools and public services, live in close proximity to big polluters, navigate criminalization and policing, and are increasingly being pushed out of their homes. Both RYSE and APEN have a long history of organizing Richmond community members, centering youth voices, and building power in Richmond. Our latest collective effort is to embed youth-led climate resilience approaches at RYSE Commons, a 45,000-square-foot resilience and liberation hub for art, healing, and community transformation. In many ways, RYSE was a "resilience hub" long before the term was popularized.

APEN youth have been learning and surveying their peers to build climate resilience into RYSE Commons. The climate resilience and liberation hub at RYSE will have a community solar and storage system to power ongoing electricity needs, power relief efforts, and provide respite from refinery pollution, wildfire smoke, and heat waves. In times of disaster, it will be a space for young people to come together to plan for and support their communities.

Has climate resiliency work allowed APEN to collaborate with other communities, like organized labor and workers?
Raval: APEN's *Resilience Before Disaster* report was jointly developed with SEIU [Service Employees International Union] California and BlueGreen Alliance and started to lay out what it means to ensure that people and neighborhoods are equipped to respond to crises. SEIU is thinking about home care workers and other caregivers and how they have an important role to play when it comes to supporting elders and people with disabilities in times of disaster. We also share an interest in rebuilding the public sector to make sure that communities are resourced and ready to meet the challenges of climate change. BlueGreen Alliance is thinking about physical infrastructure like resilience hubs,

community solar, healthy homes. These investments lend themselves to partnerships with unions because building resilience hubs can create high-road job opportunities in infrastructure building and service delivery if we do it right, and we want to see these investments bring along good union jobs that are also accessible to our community.

Our work around healthy homes and community resilience hubs has also opened up opportunities for collaboration across environmental justice groups in different parts of the state. There's no one-size-fits-all model. Leadership Council is a group that organizes in the Central Valley in Fresno, in the Eastern Coachella Valley, and they are thinking about developing a community center as a resilience hub. The needs in rural communities are not just retrofits of existing infrastructure, which we've talked a lot about in urban areas. They actually need new construction. And so that was a unique lens they brought in.

What are some of the lessons that you've learned in doing this work about what it takes to have lasting collaborations that are not just transactional?

Raval: It's just the deeper, longer-term work of trust building and being in dialogue and debate. There's no quick approach to that. Really, it takes time to cultivate those relationships and be in that dialogue. The Green New Deal coalition is positioned to build that broader tent, that bigger "we," that brings community advocates and organized labor and frontline workers to one table to vision together. That is the movement infrastructure work that feels really important to sustaining these relationships. But it isn't easy either, especially with staff transitions or turnover. Once that short-term opportunity is gone, what keeps us bound together in our fight? It's still something I think we're figuring out, but just the shared space to talk has been helpful.

And why is it even important to maintain those shared spaces and to have that collaborative relationship long term?

Raval: Well, I mean, our struggles are inextricably tied. We're not necessarily talking about two separate forces. There are different organizations, but our communities are workers, and our communities are experiencing climate impacts and environmental injustice alongside growing economic inequality. That divide of jobs/environment or communities/workers is one that's constructed to divide our power. You can't move one without the other.

10

GOOD JOBS, CLEAN AIR

HOW COMMUNITY, ENVIRONMENTAL JUSTICE, AND LABOR GROUPS STOPPED AMAZON'S AIR HUB IN NEW JERSEY

Sara Cullinane and Wynnie-Fred Hinds

Nayeli Sulca, a young woman from Elizabeth, New Jersey, started working at Amazon in 2021. Because of pandemic demand, Amazon was growing at an unprecedented rate—more so in New Jersey than in almost any other place in the world. A college student, Nayeli worked with her mother at the Elizabeth Amazon delivery station, sorting and preparing packages for final delivery. As a seasonal employee, she didn't have access to health care and was uninsured.

One day, Nayeli noticed a package was open. When she picked it up, white powder spilled out onto her arms. She started getting hives, her throat swelled up, and she felt short of breath. She asked her manager for medical attention, but he refused to send her to the hospital. She ended up calling her own ambulance, at the cost of $1,000.

Nayeli's mother suffered permanent damage to her back from her job at Amazon, where she pushed a cart with heavy packages to be sorted for delivery. After working at Amazon for a year, she had such crippling back pain that she could no longer stand on her feet for long periods of time and had lost feeling in her arm. She continued to work because she could not afford to stay home. Amazon gave her only a day's rest to recover. Eventually, her doctor advised her to quit her job. She is now permanently injured and suffers pain when she stands on her feet for too long.

Nayeli and her mother are just two of tens of thousands of Amazon workers across New Jersey. During the pandemic, Amazon nearly doubled in size in New Jersey to become the state's largest private sector employer.[1] New Jersey is a key artery in global commerce. Most products that enter the East Coast come through New Jersey—by air at Newark Liberty International Airport, by boat at the massive ports of Newark, Bayonne, and Elizabeth, or in trucks up and down I-95. It's precisely because of New Jersey's location and dense population that, over the past five years, Amazon has zeroed in on the Garden State as it expands its e-commerce empire. From 2013 to 2020, the number of Amazon facilities in New Jersey ballooned, and the company's workforce increased to 49,000 workers. As of 2020, 1 in 100 New Jersey workers was employed by the company. [2] New Jersey's elected officials have rolled out the red carpet for the mega-corporation. In 2018, in an attempt to lure Amazon's HQ2 to Newark, city representatives sent a valentine to Jeff Bezos, with the state promising $5 billion in subsidies.[3] Three years earlier, the New Jersey Economic Development Authority had awarded tens of millions of dollars in tax breaks to Amazon to build warehouses in the state.[4]

But Amazon's growth in New Jersey has been anything but positive for working-class people. A 2022 report showed that workplace injuries—in particular serious injuries, for which workers missed work or were placed on restricted duty—occurred at New Jersey Amazon facilities at a rate twice as high as other warehouses in New Jersey.[5] In 2021, Amazon injuries constituted

57 percent of all serious injuries reported in the state.[6] The negative impact of Amazon's expansion hasn't been felt just by workers. Warehouse sprawl has become a dominant concern for New Jerseyans of all political stripes. Increased traffic and air pollution as well as declining sales among local merchants have affected residents across the Garden State.[7]

Organizers and community members have pushed back against the tech giant's growing dominance. New Jersey immigrant groups have protested Amazon's ties with Immigration and Customs Enforcement (ICE), workers have fought back against unsafe working conditions and have walked off the job over wages and workplace rights, and environmental and community groups have conducted air quality monitoring surrounding Amazon facilities to measure the impact of increased truck traffic on Black and Latinx communities.[8] Residents of small towns in New Jersey have also organized to block warehouse expansion, citing increased traffic and pollution.[9]

So community groups were prepared when, in September 2021, newspapers reported a proposed backroom deal between Amazon and the Port Authority of New York and New Jersey to build a massive air hub at the Newark International Liberty Airport—a move that would have cemented the e-commerce giant's chokehold on New Jersey and the East Coast.[10] The proposed 250,000-square-foot development would have displaced current unionized workers at the airport and increased truck and plane traffic in Newark and Elizabeth, cities with frighteningly high levels of childhood asthma.[11] The Port Authority unanimously approved the deal without even listing the proposal on the public agenda or taking public comment.[12] It was a secret deal that would have harmed workers, the community, and the environment.

With only two months until Amazon and the Port Authority were set to sign the lease, community leaders in Newark and Elizabeth—the neighborhoods surrounding the Newark airport—pulled together a rapid response meeting. Among the organizations present were Make the Road New Jersey, a membership-based immigrants' and workers' rights organization, where Sara serves as executive director; environmental justice powerhouse Weequahic Park Association, where Wynnie-Fred serves as director; South Ward Environmental Alliance, where Wynnie-Fred volunteers as an activist; Ironbound Community Corporation; Clean Water Action; and New Jersey Policy Perspective. Labor was also part of the emerging coalition: the Teamsters Joint Council 73, the Retail, Wholesale and Department Store Union (RWDSU), and the Laundry, Distribution and Food Service Joint Board—labor unions that represent trucking, warehouse, and logistics workers in New Jersey—joined the coalition because they were concerned about the impact of Amazon's race-to-the-bottom business model on their industries. Rounding out the coalition was Athena, a national alliance of groups fighting Amazon's monopoly, which provided key tactical support and helped to draw connections with other Amazon site fights.

In October 2021, with airplanes roaring overhead, we stood in the middle of the vast Weequahic Park—the "lungs of Newark"—to launch Good Jobs, Clean Air NJ to fight Amazon's secret deal with the Port Authority.[13] Ten months later, after intense pressure and a powerful campaign led by local residents and labor unions, Amazon and the Port Authority of New York and New Jersey announced that they were breaking plans to go ahead with the airport deal.[14]

How did we win? Our success as a coalition depended on three key strategies. First, we built a powerful alliance based on trust and mutual interest among labor, environmental, and community organizations. Second, we successfully fought Amazon's well-financed PR machine and lobbyists to tell the true story of Amazon's impact on New Jersey communities and workers. Third, we polarized elected officials and key decision-makers by making it clear that standing with workers and communities meant opposing the air hub. Our victory shows that when community groups, labor, and environmental organizations join together to build power, we can take on the largest, most powerful companies in the world and deliver justice for our communities.

Building a Powerful Alliance Among Labor, Environmental, and Community Groups

The coalition was the first time that most of the labor, community, and environmental groups had worked together, despite having been active in neighboring communities for decades. It was challenging to build trust and a shared agenda among organizations with divergent priorities over a very short period of time and under intense pressure. We did so through weekly meetings, first over video conferencing and then in person. We began with the premise that alone we could not defeat Amazon: we had to build a strong coalition in which partners committed to fighting for one another's priorities. That meant not caving if Amazon tried to greenwash the deal by providing funding for electric trucks or if labor unions were able to win a labor peace agreement. We created a list of shared demands that included mandatory community-impact meetings, noise and traffic studies, enforceable benchmarks for zero emissions at the air hub, funding for small businesses and community groups, a labor peace agreement, and enforceable labor standards. We decided that we would not accept any deal unless Amazon and the Port Authority agreed to all of our demands.

Carrying out the work of the campaign together gave us an opportunity to deepen our relationships and to get to know one another on a personal level. One important moment came during a training for a community canvass. The coalition decided to launch a pledge card drive in the neighborhoods surrounding the airport. Our goal was to collect four thousand petition signatures and have thousands of one-on-one conversations about the air hub. Every week, dozens of members of immigrant rights groups, environmental justice organizations, and unions joined with Amazon workers to speak with community residents about the threat the air hub posed. We used a deep-canvassing approach, in which our members engaged residents in one-on-one discussions about issues meaningful to them.

One of the volunteers for the community canvass was Teamsters member Tyrone Gilliam. His daughter had worked for Amazon and had experienced unsafe working conditions while she was pregnant. He saw the sharp difference between his daughter's non-union Amazon job and his union position. He also lived in Elizabeth, close to where the air hub would be built. Tyrone's story was not unique. As the training began, each community and union member shared stories like Tyrone's—stories about employment-related injuries, children with asthma, or the lack of quality work as the corporation came to dominate the labor market. It became harder to see division. Everyone was affected by Amazon, and many were affected in multiple ways—as a worker or future worker, as a member of the community, or as the parent of a child breathing polluted air. A union contract wouldn't solve the problems of pollution or increased truck traffic that workers living in nearby neighborhoods would face. And a commitment to clean energy at the air hub wouldn't fix the backbreaking injuries. Everyone was going to bear the brunt of the air hub, and everyone had to fight it—together.

Fighting Amazon's Well-Financed PR Machine to Tell the True Story

When we began the campaign, we heard a common refrain: *Amazon may be a union-busting monopoly, but Amazon jobs pay better wages and offer health care, and that's not something that Elizabeth and Newark can walk away from.* Amazon had spent millions on billboards up and down the state advertising signing bonuses and flexible hours. The company gave strategic donations to elected officials and nonprofits in Elizabeth and Newark to sweeten its entry into the communities. The truth, of course, is that Amazon jobs are dead-end and dangerous. Our task as a coalition was to bring to light the true impact of Amazon on workers and communities.

We combated Amazon's misinformation campaign by building platforms for Amazon workers to share their own stories. Current and former Amazon workers joined the coalition and told the public about the impact of Amazon jobs on their lives. Christian Rodriguez, a former Amazon worker and an environmental justice advocate with Ironbound Community Corporation, relayed their story of overwork and surveillance at Amazon at a press conference before a Port Authority meeting. "When I worked at Amazon," Christian told the press, "I was monitored like a robot and overworked to increase the profits of the richest man in the world. We would be given such short bathroom breaks and had to walk so far across the warehouse that by the time we got back we barely had time to go to the bathroom—we had to run back. If our production failed, we had to speak to a supervisor about our numbers and figure out a way to stay on track."[15] Workers like Nayeli Sulca and her mother were already feeling the effects of Amazon's unsafe working conditions, and with the air hub set to open soon, they worried about the company's negative impact on the air they breathed at home too.

The Good Jobs, Clean Air NJ coalition worked with researchers to conduct studies of Amazon's workplace practices, including injury rates, wages, and turnover rates. Researchers from New Jersey Policy Perspective and Rutgers University analyzed OSHA injury logs and found that injury rates among Amazon workers were skyrocketing in New Jersey—and that Amazon accounted for 57.2 percent of all serious injuries in the general warehousing and storage industry in New Jersey.[16] The injury rates illustrated in stark terms that Amazon jobs are backbreaking and unsafe.

What's more, wages across the industry declined during Amazon's rapid growth in New Jersey. Between 2015 and 2020, Amazon's New Jersey workforce grew by nearly 800 percent. In that same period, New Jersey's delivery and courier sector wages declined 10 percent when adjusted for inflation; warehousing and storage wages declined 17 percent when adjusted for inflation.[17] Turnover at Amazon was also exceedingly high. Researchers from the National Employment Law Project found that annual worker turnover at Amazon warehouses in New Jersey is approximately 124 percent, almost double the rate of turnover at non-Amazon warehouses in the state.[18]

The coalition worked with researchers to release each report just before the Port Authority held its monthly meetings. The press widely covered the reports, and our coalition members upped the pressure by rallying in front of the Port Authority meetings and testifying about the impact of Amazon's bad jobs on workers, the economy, and the environment.[19] We packed the hearing rooms to make clear that there was broad support for ending the deal with Amazon. The results were powerful. New Jersey's largest newspaper wrote an editorial urging an investigation of Amazon's injury rates.[20] In the spring, the New Jersey General Assembly called a special hearing on Amazon's injury rates, and OSHA launched an investigation after a member of the U.S. Congress called for increased oversight.[21] Public opinion on Amazon was beginning to shift in New Jersey.

Drawing a Clear Line in the Sand to Polarize Elected Officials

Around the same time, Amazon workers on Staten Island and Bessemer, Alabama, were organizing to win a union and facing extreme retaliation. In a pro-union state like New Jersey, we saw an opportunity to push elected officials who supported workers' rights and who cared about the environment to back our campaign to hold the Port Authority and Amazon accountable for the secret deal at the air hub. Our coalition reached out to local elected officials from the municipalities surrounding the airport and urged them to speak out against the deal. We held multiple town halls at which local officials like Mayor Dahlia Vertreese of Hillside, Mayor Christian Bollwage of Elizabeth, and State Senator Joe Cryan, who represents parts of Union County, heard directly from workers and community members about the impact the air hub would have on our communities. After meeting with our coalition and local residents, Mayor Bollwage of Elizabeth and Mayor Ras Baraka of Newark—previous Amazon boosters—publicly voiced concern over the deal and urged Amazon to sit down with community leaders.[22]

We brought the same public pressure to the monthly Port Authority meetings. After months of knocking on doors, attending local town halls, and organizing citywide meetings, in the spring of 2022, local residents packed the Port Authority board meetings, providing hours of testimony to oppose the air hub. Coalition members also delivered thousands of petitions to the Port Authority demanding an end to the secret deal. In a particularly powerful moment in the campaign, community leader James Young, a resident of the South Ward of Newark and a member of the South Ward Environmental Alliance, shared the story of a nine-year-old neighbor who died of an asthma attack. "The air quality is already a problem for our community. . . . What are you doing to ensure the health of our children?" James asked the board.[23] As James spoke, crowds rallying downstairs could be heard through the boardroom walls.

After the Amazon Labor Union's historic victory at the JFK8 Amazon facility in Staten Island—just minutes from Elizabeth and Newark—the campaign to stop the air hub picked up even more momentum. In June 2022, we organized a letter signed by more than a dozen federal, state, and local New Jersey elected officials urging the Port Authority to listen to local residents and workers.[24] After meeting with Amazon workers who were also members of our coalition, Congressman Donald Payne Jr., who represents Newark, wrote an op-ed calling on Amazon to listen to community and labor concerns about the air hub.[25] It was clear that Governor Murphy was also taking the demands of the groups into consideration.[26] Eight members of the New Jersey congressional delegation penned a letter to OSHA urging an investigation of the rapidly rising injury rates at Amazon in New Jersey.[27] In early July, the Port Authority and Amazon announced the end of the deal.[28]

What's Next

As we planned a community party in Newark's Weequahic Park to celebrate a victory of environmental justice communities and organized labor over one of the largest corporations in the world, it quickly became clear how much work lay ahead. The morning of the celebration, we learned that an Amazon worker at a Carteret, New Jersey, fulfillment center had died on the job. We soon came to know that the worker—who died while working during Amazon's Prime Day—was Rafael Mota Frias. Statements to the press from Mota Frias's coworkers indicated that extreme production pressure, high temperatures in the warehouse, and failures in Amazon's emergency response likely contributed to his death.[29] And Mota Frias was not alone: his was the first of three Amazon worker deaths in three weeks in New Jersey, between mid-July and August. One week, working-class New Jerseyans were celebrating a win over Amazon, the next, they were mourning one of their own. OSHA has since announced a comprehensive investigation of the deaths.[30]

Good Jobs, Clean Air NJ member organizations are continuing to organize to reign in Amazon's greed and ensure that communities of color and workers have a say in developments in our neighborhoods. We are meeting with environmental justice organizations in Washington, California, Tennessee, and Georgia to share what we've learned from this fight. Through the Athena coalition, worker organizers from New Jersey are also hearing from worker centers in Minnesota, Missouri, and Illinois about their experiences organizing direct actions to win concessions from Amazon.

Just a few years ago, it seemed inevitable that Amazon would slowly creep toward controlling ever more aspects of our lives. Now the tide is turning. Thanks to a coalition built on trust, common vision, and careful strategy, people power is winning.

11

KILLING THE WIINDIGO

RESTORING INTERDEPENDENCE AND UNITING OUR MOVEMENTS

Winona LaDuke and Ashley Fairbanks

We are living in a time of transition. We are reaching the end of the constant exponential growth of the industrial world, growth which has been built on the foundation of cheap energy. The fossil fuels that powered the engines of capitalism are running out, and our planet is telling us that she has had enough.

This economy, which has made a select few wealthy beyond measure, has required many sacrifices. The poor have been forced to suffer, toiling away for poverty wages. Everyone's well-being has suffered as we've traded our time, energy, and connection with one another for a paycheck that often still struggles to keep up with our needs and our desires. More than anyone else, our Mother Earth has known sacrifice—her ecosystems, her biodiversity, her very essence face mounting losses.

We call this economy, the one built on suffering, on taking and not restoring, the wiindigo economy. The wiindigo is the cannibal of Anishinaabe legend. Much like the wiindigo, our modern economy cannibalizes us.

We have had enough. But the wiindigo will not die—he must be killed.

The scale of transformation we need will require mass social movements, political will, and completely restructured systems.

This essay is about how we center Indigenous knowledges, reestablish interdependence, and plant the seeds of a new world—one that we build together. In honor of one of the greatest political leaders of North America, we choose to call this the Sitting Bull Plan.

Sitting Bull, or Tatanka Yotanka, was a Hunkpapa Lakota political and spiritual leader during the most complex times for the Oceti Sakowin, the People of the Seven Council Fires. His people saw the destruction of the buffalo, the center of their way of life, as well as the coming of the military and the Indian Wars. Sitting Bull's vision led to the defeat of George Armstrong Custer at the Battle of the Little Bighorn, and his leadership decisions have been widely regarded as some of the most brilliant in the history of war and in civil society. His vision was great, and his times called for it. These are those times as well.

Lay the Foundation

Re-centering Indigenous knowledge is the foundation of the Sitting Bull Plan. *Kohtr'elneyh* means "we remember" in Benhti Kenaga', the Lower Tanana Dené Athabascan language. The organization Native Movement first came up with the idea of using this term and acknowledging traditional knowledge as a guiding light in just transition work. "The Just Transition framework," Enei Begaye of Native Movement reminds us, "upholds relationships that stem from time immemorial, and must recognize it was not people who governed the land, but rather the land and the spiritual beings of the land prescribed relationships among humans and their relationships to the land. We acknowledge the depth of Indigenous knowledge rooted in the long inhabitation of a particular place offers answers as we search for a more satisfying and sustainable way to live."[1]

Indigenous people hold the memories of a sustainable world in the DNA. Many of us are just a few generations removed from a sustainable life lived in harmony with our natural world. Some of us are still living it. Indigenous people connect differently to the natural world because we don't divorce ourselves from it. We see ourselves as a part of the ecosystem.

In Anishinaabemowin, our language is gendered animate/inanimate rather than male/female. And what we see as animate is far more expansive than what most Euro-Americans might expect. *Mitig* (a tree). *Asin* (a stone). Living in right relationship with the natural world comes easier when it's baked into your language.

Capitalists dismiss animistic traditions like ours, traditions held by so many of the world's Indigenous peoples, because the success of the capitalist system requires that the Earth be seen as a collection of resources ripe for plunder. Now we pay the price.

Re-centering Indigenous knowledge is a critical foundation for building our new world.

Stop the Bleeding

Once we've laid the foundation of our plan, we have to stop the bleeding—bleeding caused by consumption that is pushing us closer and closer to climate catastrophe. About 70 percent of the U.S. economy is based on consumption.[2] Our need for stuff demands a lot of extraction. It requires a lot of mining and a lot of plastics. Many mining projects are so inefficient, they might as well be looking for unobtanium, the stuff of legend in the film *Avatar*.

Demand for growth and the consumption it requires have destroyed the connection between people and things—we have an easy time accepting subhuman treatment of workers when we don't have to see them. We don't see the people who toil in mines for our lithium iPhone batteries or who sew the $2 pants for sale on Shein. Being removed from this suffering allows us to consume in a way that rarely confronts us with reality. And it's only getting worse, as shopping continues to move online and Amazon warehouse workers and drivers toil under terrible conditions to seamlessly deliver us objects as fast as we order them.

The suffering is closer to us, yet somehow farther away.

This level of consumption and extraction is not sustainable, for many reasons. The climate crisis is just one. As the global population continues to grow to 8.5 billion by 2030, 1.3 billion new people will enter the global middle class.[3] Our planet cannot support 1.3 billion additional consumers. We have to push back on the wiindigo idea of endless growth and consumption that has dominated the last few centuries. We have to champion the global economy shrinking, not growing.

There's a name for the idea of shrinking our economy: degrowth. Degrowth is a critique of the capitalist push for endless growth. In its original French, *la décroissance* refers to a river going back to its normal flow after a disastrous flood. This definition fits nicely into our Sitting Bull Plan—we seek to restore the natural order of things after the flood of industrialization, globalization, and capitalism has wreaked havoc on our lands, our animal relatives, and the global population of Indigenous people.

Reshaping our economy will allow us to improve the conditions for people, because people exist as more than workers, but as mothers, children, grandparents, and community elders. When people are seen as part of our village, when we understand that their well-being is interwoven with our own, we are called to treat them with more care. When we are no longer concerned with growing economies on paper, we can focus on quality of life as a key metric, instead of a growing GDP.

We must reject the demand for growth, shrink our economies, and reconnect to the people making our things. It will be a win for our planet, and a win for people.

Plant the Seeds of the Future

They tried to bury us, but they didn't
know that we were seeds.
—Zapatista proverb

We must restore our small economies, and that includes regaining our ability to grow our own food. That work starts with our seeds. Seeds are about promise, hope, and commitment. Seeds are about life. Indigenous and heritage seeds are intelligent seeds. Modern, genetically modified seeds are not.

Major corporations like Monsanto and Syngenta use billions of dollars to create "climate smart" varieties of seeds. Patrick Mooney, a seed scholar, told an Indigenous Slow Food gathering in Meghalaya, India, that the average cost for a climate smart seed is $136 million per seed variety.[4] In the meantime, Indigenous nations worldwide are already adapting our pre-petroleum varieties for the times ahead.

Combined, Indigenous and peasant farmers produce 70 percent of the world's food.[5] Critically, Indigenous and heritage plants are not addicted to industrial agriculture, to toxic cycles of fertilizer and pesticides—chemicals that don't just harm the ecosystems where they are used but also harm our bodies and the bodies of our winged, hooved, and scaled relatives. Organic and restorative agriculture sequesters carbon, rebuilds topsoil, and cleans water. It's estimated that organic agriculture on a worldwide scale could sequester up to a quarter of the carbon in greenhouse gasses today.

Today, 15–20 percent of total annual greenhouse gas emissions is attributed to agriculture, much of that from plowing.[6] Carbon can be put back into the soil by no-till farming practices, perennialized agriculture, the application of compost and manure, and the holistic grazing of livestock, like the buffalo commons and chickens. This is how life is restored. Feed the topsoil and feed the Earth.

Cultivate Hemp, the Wonder Fiber

Indigenous cultural predilection for systems-based thinking allows us to see interconnectedness in a unique way. This thinking allows us to reverse engineer solutions. One solution that has presented itself in recent years is the use of the cannabis plant. Cannabis has the ability to transform our economy, return carbon to the soil, and be a central material in sustainable housing and heating. It can provide health and well-being and a path to restorative justice. This potential applies to both the plants known as industrial hemp and marijuana. In a world of technological fixes, it turns out that the future should have more plants in it, not less.

Textile production is one of the world's most polluting industries, producing 1.3 billion tons of carbon dioxide equivalent.[7] The textile industry used 98 million tons of oil in 2015. As Valerie Vande Panne writes in *Salon,* "By 2050, that number is expected to be 300 million."[8] Our resources are too precious for that kind of waste.

Hemp is the solution. It has about three times the tensile strength of cotton. It is mold and UV resistant and uses very little water, pesticides, or fertilizers. Hemp produces twice as much fiber per acre as cotton does. Not only can hemp replace cotton, it can replace one of the other large burdens on our planet—plastic. Hemp has the ability to do basically anything you can do with plastic. But hemp plastics biodegrade.

Tribal nations hold large land areas, so there is great potential for tribes to become hemp growers. In 2020, we provided hemp seeds to farmers from five tribes in our region, all of them interested in the hemp materials economy. Tribal farmers from the Rosebud, Cheyenne River, Navajo, Red Lake, and Oneida nations were all given seeds from our stocks. In addition to seeds, we are providing support to tribes interested in developing their own hemp policies.

Along with hemp, the Indigenous cannabis industry has already begun to blossom, from the NuWu Paiute dispensary in Las Vegas, Nevada, to the Indigenous hemp collaboratives evolving in the Northern Plains. As tribes begin to expand into the cannabis industry, it's critical to also flex our tribal sovereignty to address the injustices of the drug war, which have impacted Indigenous people at a disproportionate rate.

Choose the Green Path

Of course, something has to power this next world. Thankfully, we've got an unlimited resource just 91 million miles away. According to Anishinaabe prophecies, we are in the time of the Seventh Fire. At this time, it is said we have a choice between a path that is well worn and scorched and a path that is green and unworn. If we move toward the green path, the Eighth Fire will be lit and people will come together to make a better future.

We choose the green path. Our company, 8th Fire Solar, builds a sustainable product called solar thermal panels, a relatively simple technology that turns sunlight into heat. Most American households, especially those in tribal communities, spend the majority of their home energy on heating. A single solar thermal panel significantly reduces a home's fossil fuel use and carbon emissions.[9]

In addition to providing direct energy savings, 8th Fire is built on investing heavily in building up the local workforce, training tribal citizens at White Earth in careers that are rewarding and that pay the bills. 8th Fire employees build each panel and are able to teach others how to build them.

This is the kind of solution our new economy needs, solutions that solve multiple problems at once. It's possible to provide good jobs, save people money on energy bills, and reduce our dependence on fossil fuels. We're doing it, and we're not the only ones.

Across Turtle Island, solar projects are popping up. Red Cloud Renewable in Pine Ridge, South Dakota, is empowering and educating tribal members to work with solar power and to help their tribal communities become energy independent. On the Northern Cheyenne Reservation, the White River Community Solar Project is training tribal members and building a 1.25-megawatt solar project. Solar Bear is leading the construction of a 13-megawatt solar farm right down the road in Red Lake, Minnesota.

For too long, the well-being of workers has been pitted against our planet, when those things are fundamentally connected. To move forward, we have to show people that they cannot thrive while the land suffers. We have to show them how we are connected and that they will be taken care of if they take care of the land.

Take the Green Path

Now that the climate crisis is here, we have a choice. We can continue to buy into the idea that we must consume, grow our economies, exploit our fellow humans, and extract all life from our only home, our mother, *omaa akiing*.

Or we can choose the green path. We can choose the path of love, empathy, and interdependence with all living things.

We don't see a climate movement, a labor movement, a civil rights movement. We see the wiindigo economy, and all of us. Will we work together to kill it, or will we—and all life on Earth—be its next victim?

Now is the time to think intergenerationally, to think deeply about justice and peace and our stewardship of this planet and all the beings that live upon it. This is our gift to our descendants.

Now is the time to dream future generations into existence. To imagine a better life for them and to dwell in a place flowered with optimism and grounded in action.

Now is the time for us to plan and enact our just transition to the next world, a place where all creatures live with dignity, where our streams run clear and the stars shine bright.

Now is the time to kill the wiindigo, to burn down the extraction economy like a prairie, so something new, something better, can grow. Let's light the Fire.

Part III

"WE ARE OUR BEST CHANCE FOR RESCUE"

ORGANIZING THE FRONT LINES OF THE CLIMATE CRISIS

12

"WE ARE OUR BEST CHANCE FOR RESCUE"

GREEN WORKERS ORGANIZING FOR MORE AND BETTER JOBS

Matthew Mayers

> I witnessed several co-workers fall . . . from heat exhaustion and lack of water. Every day someone leaves work because they are dizzy and exhausted. It has been extremely hot outside and we have no access to shade tents, umbrellas, or any place to get out of the sun and cool off.

This is not a report from a farmworker or day laborer. It is testimony from the front lines of America's clean energy revolution, submitted by a laborer at Bighorn Solar 1, a utility-scale solar farm installation in Pueblo, Colorado.[1]

For all the glowing talk of a green jobs revolution, the reality for clean energy workers is low pay, dangerous conditions, and precarious employment. Too often, high-paying, unionized fossil fuel jobs are being replaced by low-road renewable energy jobs. As a result, we are not only missing the opportunity to build a more just, sustainable economy but also losing the chance to win the support of much of the working class for this transition. Only by empowering the renewable energy workforce to fight for—and win—good green jobs will we achieve a just transition to a clean economy.

This is the challenge I decided to take up when I co-founded the Green Workers Alliance (GWA) in October 2020. We knew that we needed to build an organization to mobilize green workers, the vast majority of whom are not yet unionized. We also knew that solar installers and wind turbine technicians were two of the three fastest-growing jobs in America.[2] With unionization rates in renewable energy still low (estimated to be around 10 percent overall, though higher in some parts of the sector), we need creative approaches to build organization and to strengthen the ability of workers to form unions.[3] At GWA, we wanted to experiment with new methods to reach green workers, learning from the success of online-to-offline campaigns, like those United for Respect had pioneered in the retail industry. And we wanted something that could improve conditions for non-union workers, even as we helped create a path to unionization for interested workers.

GWA is building a worker organization made up of current and aspiring renewable energy workers demanding a just transition from fossil fuels and a massive investment into green energy. We are uplifting the voices of renewable energy workers to make sure a just transition prioritizes more green jobs, better pay and benefits, and improved working conditions in the renewable sector. GWA is helping workers navigate the critical and fast-growing renewable energy industry through community building, leadership development, and trainings. Our goal is to organize tens of thousands of renewable energy workers and to harness worker power to hold the utility sector and renewable energy industry accountable for creating more and better green jobs that benefit both workers and the planet.

This piece describes some of GWA's experiences trying to build power in the sector as we organize workers and lay the ground work for the struggle with the investor-owned utilities that ultimately hold power in the industry.

The Utility-Scale Renewable Energy Industry

There is an urgent need for worker organization in the renewable energy sector. The industry is dominated by temp firms and staffing agencies, meaning green workers are constantly searching for the next gig, with no job security, poor benefits, and dangerous working conditions.

A study from the Political Economy Research Institute at the University of Massachusetts Amherst showed that the Inflation Reduction Act would create more than five million clean energy jobs over the next decade.[4] Yet the poor conditions in the industry make it hard to attract and retain the growing workforce this sector needs. Many workers later leave the field—sometimes going back to the very jobs in the fossil fuel industry we are trying to replace because they offer more secure working conditions.

The issues these workers face are legion. Pay is often low, nepotism and wage theft are common, benefits are scarce, layoffs are frequent, and safety problems are widespread. Promised per diems often arrive late, if at all, leaving workers to sleep in tents or in their cars for many days. Stories like those of GWA member and solar installer Alicia Ramirez are all too common. Alicia traveled from Texas to Georgia on her own dime for a promised project, only to find out that none of the twenty people who traveled to the site would be hired.[5] A piece in *Vice* featured solar installer Thomas Shade describing the industry this way: "You're always away from friends and family. Sometimes you don't know anybody. . . . They don't want to pay you enough for your room and for you to eat for the week. . . . So you got two guys in beds and a guy sleeping on the floor, one guy on the couch or a chair."[6]

Such poor conditions grow out of the industry's reliance on temp firms and staffing agencies, as general contractors compete in a race to the bottom to achieve cost savings. Workers often do the same job for very different pay, depending on what temp firm they work for or what they negotiate with a labor recruiter. At the same time, the investor-owned utility companies that buy the electricity from these projects are raking in astronomical guaranteed profits.

As wind technician and trainer Trent Nylander argues, corporate greed creates incentives for understaffing and overworking inexperienced subcontractors.[7] Meanwhile, the lack of OSHA regulations means there is no federal oversight.

> It is well known in the industry that OSHA has had no guidelines when it comes to working in the wind industry. Knowingly, the fact that OSHA's guidelines and protocols stop at the door leave a lot of safety systems bypassed or in such ill repair. The wind industry isn't a new industry, and many of the first generations of turbines have been pushed beyond their twenty-year lifecycle. And yet we are continuing to climb these huge structures with our lives in our own hands, with no oversight from OSHA.
>
> The fact that OSHA has not trained a single inspector on these three-hundred-feet-high power plants since their inception is astonishing to say the least. When an incident occurs in a turbine, we are our best chance for rescue. Emergency responders have to wait on us to reach the bottom of the tower, as they too are untrained in rescue at heights.

These conditions will change only if workers organize to build power. That is what we set out to do in the fall of 2020.

GWA Organizing Tactics

We built the GWA at the height of the COVID-19 pandemic, which forced us to take an online-first approach. We searched for workers online, running experiments with Facebook ads before turning to Facebook mining—a process that allowed us to find workers with workplace issues or concerns in online groups.

Starting with a very broad definition of green jobs, we quickly found a base among utility-scale renewable energy workers. These workers constantly travel for gigs on large solar or wind sites, usually in rural areas. They are also frequently online, looking for the next job and surfing huge digital groups like American Solar Farms, which has over seventeen thousand members. Facebook proved the most important place to be; it is where workers of all ages go to find jobs and build connections in the field. We became close with the moderators of some of these groups, and workers began to direct their coworkers to us when they had workplace problems. Like worker centers—but virtually—we began to help people with workplace issues and to recruit those willing to stand up for workers' rights. As GWA leader and solar worker Brittney Linton said in her remarks at our first national meeting,

> I became a part of GWA, because [the organizer] here reached out to me. . . . I was online on the American Solar Farms page, and someone had an issue, and I put my two cents in. "Hey, you have your rights." Coming from the corporate world, I knew my rights as an employee, and a lot of people don't, and I said, "You should do this, you should do that," and [the organizer] reached out to me, so I've been with GWA for like a year and a half.[8]

But workers also began to join because they were looking for fellowship and connections in a field where they were frequently traveling and working with new people. Even though the field is large—with perhaps one hundred thousand workers toiling on utility-scale sites—we found that many of them knew one another from years of cycling through different projects.

As time went on, we began other digital experiments to bring members into the organization, from workers' rights trainings to online trainings on job hunting. While these trainings didn't end up bringing many members into the organization, they do give us credibility as an organization concerned with improving workers' positions in the field, and they remain on our website for people to access. In the spring of 2022, we also began to do in-person organizing, hitting job fairs and having GWA members introduce organizers to workers near job sites. Much of this work was done by one or two organizers, with members volunteering to run most of the workers' rights and job hunting trainings.

We are now up to 1,100 members and rapidly growing. The bulk of our growth has come from connecting with people online or in-person and having organizing conversations about the issues they face and why building a strong organization of workers could improve conditions. After we held our first in-person meeting of fourteen GWA leaders in September 2022, members became increasingly active in recruiting coworkers to join the GWA Facebook group, which continues to be the primary space where members can connect with one another and learn about GWA activities. Given the dispersed nature of this workforce, online activities like our monthly leadership meetings will continue to be key.

Much of this growth is because of an exceptional lead organizer who recruited and maintained connections with workplace leaders. They brought their experience as a field organizer in innovative labor campaigns like Fight for $15 and adapted it to a largely online effort. At the same time, it has been hard to train up newer organizers to do this work, especially in a remote and now hybrid work environment. We are experimenting with bringing more workers off the job to do organizing, something that is made easier by the intermittent nature of their employment. We will continue experimenting with new methods to grow our membership and develop their leadership to build power in this industry.

Challenges and Opportunities with This Workforce

The majority of the renewable energy field is still made up of white men, but that is changing, and the membership of GWA is much more diverse than the field—particularly at the leadership level. This is partly because women and people of color face the most discrimination and see the need for change most clearly. It's been exciting to see them take the lead.

Our organization's leadership is diverse not only in terms of background but also in terms of politics. This means we have to accept that our members' views may differ from our own even as we work to make progressive change. What does that mean in practice? It means we stay out of partisan politics, but we do talk about the impact of elections on renewable energy rollout. During the COVID pandemic, it meant that we accepted that many of our members are resistant to vaccination and masking, even as we worked to take on misconceptions around the issue through respectful discussions with members who decided to get vaccinated. At the same time, we have no tolerance for discrimination or hate, and we challenge those who spew it on online message boards in the renewable energy industry. Striking the right balance between accepting members' beliefs and holding the line on core values is vitally important because organizing this workforce—and other workforces like it—is crucial not only to winning better working conditions but also to building durable progressive majorities.

The arc of GWA leader Cat McCoy is instructive. Cat comes from coal country and was a Trump voter in 2016. A woman in the male-dominated field of heavy equipment operators, she came to GWA for help with a sexual harassment case on the job. After GWA helped get her case resolved, she became an active member and recruited others to join. She came with GWA to COP26 in Glasgow as part of a larger coalition called It Takes Roots. This was her first trip out of the country, and it was an amazing experience for her, as she protested with Indigenous leaders and met frontline climate activists. She wrote about her story in the *Earth Island Journal*.

> My whole life—from my work at the uranium mill to my time in the scrap recycling industry—I have seen first-hand the ravages of our environmental destruction. I am grateful that over the past six years I have gotten to play some role in decreasing our human footprint on the planet. Now I join my colleagues in demanding that our national leaders remember that clean energy is built by those of us willing to climb a wind tower, or spend hours outdoors in poor weather installing solar panels, or climb someone's roof. We deserve to have the resources to take care of our families as we care for the planet.[9]

Not every GWA leader is going to go to a COP summit, of course. Our belief is that workers' attitudes will change as they engage in struggle alongside progressive allies. Along those lines, we've built coalitions with frontline groups and progressive climate organizations, several of whom have met with GWA members.

Campaigns Targeting Utility Corporations

Our struggle must be with those who have the power to make real change in the renewable energy sector: the investor-owned utility companies that have the power to force contractors to improve conditions on renewable energy job sites. At the same time, utility companies are often laggards in adopting renewable energy, even as they account for 25 percent of the nation's climate emissions.[10] Recent Supreme Court decisions, along with Joe Manchin's weakening of climate legislation, have made it much harder for the federal government to force them to do better. And, of course, these utilities are taking advantage of customers with rate hikes and shutoffs, a problem that will only get worse with climate change.

GWA is working on campaigns to make substantive demands on utility companies to quickly transition from fossil fuels, build renewables at scale, ensure that workers can form unions and receive pay and benefits commensurate with fossil fuel jobs, and end rate hikes and shutoffs of moderate-income customers. These efforts will be driven by a coalition that includes unions, climate groups, and community organizations.

In addition to mobilizing workers in the field, we are using a wide array of tools and tactics, from shareholder activism to digital campaigns among utility consumers. Some of these campaigns are led by GWA, while in others, we are partnering with exciting efforts like the Power 4 Southern People NOT Southern Company—an effort led by local groups in Alabama, Georgia, and Mississippi in conjunction with the Arm in Arm project of the US Climate Action Network.

With the room for federal political action becoming much more limited, it is through campaigns like this, which target corporate actors and put workers and frontline communities at the center of the fight, that we will make progress on fighting climate change and winning economic justice. This is particularly the case in red states, where progress at the state level is often impossible and where even local action is often forbidden by state preemption laws.

Conclusion

GWA members are proud of the work they do. As wind turbine installer Joe Zimsen said in an op-ed in the *Cedar Rapids Gazette*,

> I work to keep our country safer, cleaner, and empowered. I'm proud of my work helping to end our dependence on volatile fossil fuels and building real American energy security. There is so much more opportunity to grow this field and create thousands of fulfilling and good paying jobs.[11]

There are more immediate reasons workers are staying in the field. As solar worker Joe McCoy said after returning to the field from a year's absence, "I came back to solar because I enjoy the connection you make with the people on crews you work with and the different aspects of traveling around the country."[12]

At the same time, these workers know we need to improve jobs in this field so they can better support their families and so that the industry attracts and keeps the best people in the sector. Wind technician Patrick Foeday, who has been active with GWA from the beginning, put it best when he said that we must organize "to fight for better pay and benefits, improve the safety and well-being of technicians in the field and win equal opportunity regardless of your gender or race."[13]

We need more green jobs, but we need better ones too. This is going to require organizing among the workers at the front lines of this transition. It is going to require holding corporations accountable to share their massive profits with the workers making the clean energy revolution possible. And it is going to require strong alliances among worker organizations, climate activists, and community organizations. The challenges are daunting, and there is much more to be done, but we know that, together, we can build a movement that wins economic justice and a green economy.

13

SOLIDARITY FOR THE SNOWPACK

Isabel Aries, Ryan Dineen, and Katie Romich

In the Mountain West, climate change is causing the disappearance of winter snowpack, posing an existential threat for winter sports workers and local hospitality workers. Simultaneously, ski patrollers and other outdoor recreation industry workers are forming unions to fight for better pay and safer worker conditions and to elevate their work as a profession. The intersection of these two trends creates a unique opportunity for the growing labor-environment movement in the West.

As seasonal laborers who take other jobs in the offseason, ski industry workers face significant challenges in unionizing. They rely on local community allies for support as they knit together a national outdoor recreation workers' movement. Uniting with the environmental movement would bring legitimacy and resources to their fight for better pay and working conditions—in addition to the shared goal of saving winter.

Winter sports workers are strategically important for the environmental industry as well. They are rank-and-file union members who challenge the narrative that jobs and climate are in opposition. For ski industry workers, stopping climate change will protect jobs.

Snow is disappearing—and these workers' jobs will disappear

with it. To protect their communities and the future of their work, patrollers and others must do more than struggle with employers on bread-and-butter issues. They must also get active in the fight against climate change, providing a strategic opening for labor and environmental justice groups to forge coalitions that can strengthen both movements.

Disappearing Snowpack Threatens Workers in Mountain Communities

In mountain towns across the country, the lead-up to the ski season brings cold nights, changing leaves, and the ever-present roar of snowmaking machines. The white noise stokes excitement as communities look toward the coming winter. This is the modern version of the anticipation caused by the first snowfall of the year.

Ski communities transform in winter as the snow, both real and artificial, brings visitors and enlivens local economies. In Colorado, winter sports anchor the tourism industry—the state's second-largest industry after oil and gas. The state's 2022 revenue from tourism was $26.9 billion, compared with $37.4 billion in oil and gas.[1] The tourism industry drives job creation in Colorado. The Colorado Office of Economic Development and International Trade (OEDIT) reports that in 2016, Colorado had 1.7 percent of the U.S. population, but 7.7 percent of the nation's tourism jobs.[2] The industry generates more job growth than does any other industry in the state.[3] Tourism also bolsters the state and local tax base; according to OEDIT, travel-generated state and local tax revenue reached approximately $670 per Colorado household in 2021.[4]

To attract visitors, Colorado must have snow. And to bring snow, ski resorts increasingly rely on snow machines. First introduced in 1952, snow guns have permeated the ski industry since the 1970s. Today, 90 percent of ski areas depend on snow guns to guarantee snow for the duration of the ski season. Snowpack measured in April has declined by 20–60 percent at most monitoring sites. The problem is only getting worse. A recent study from Los Alamos National Laboratory in New Mexico suggests Colorado will experience a 50–60 percent reduction in snow by 2080.[5] This translates to shorter seasons, more dangerous avalanche conditions, and an increased reliance on artificial snowmaking.

A painful irony of this increased reliance on artificial snowmaking is that it requires massive amounts of water and energy—thus contributing to climate change, which, in turn, exacerbates the precarity of natural snowfall. One analysis estimates that ski resorts use about 1.5 billion gallons of water per year in Colorado alone.[6]

Diminishing snowpack will likely lead to increased drought and forest fire risk, lower property values, agricultural losses, and economic insecurity. It also means that workers in ski communities will lose their jobs. Already, climate change is threatening ski patrollers' work. The problem will only get worse, which is why ski patrollers across the West are beginning to turn to their unions to protect winter as a workplace issue. This move comes during an upsurge in unionization in the ski industry.

Upsurge of Unionization in the Ski Industry

Ski industry workers are organizing for higher wages, job security, safer working conditions, better training, and allowances for gear, such as gloves to keep their hands warm. And they're winning.

During the 2020/21 ski season, the ski patrollers of the Breckenridge Ski Resort successfully self-organized into the United Professional Ski Patrols of America/Communications Workers of America (UPSPA/CWA) Local 7781. They followed patrollers at Big Sky Ski Resort in Montana, who voted 69 to 21 to unionize the same season, as well as existing union patrols in CWA Local 7781 at Park City in Utah, Stevens Pass in Washington, and Crested Butte, Steamboat Springs, and Telluride in Colorado.[7] The next season, professional patrollers at Purgatory Resort, in Durango, Colorado, voted 35 to 3 to unionize. "Right out of the gate, we had pretty high support," Purgatory patroller Jason Moore said. "We thought it would be worth organizing and giving ourselves a better opportunity for a collective voice."[8]

At the same time, patrollers at the Park City Mountain Resort in Utah prepared to go on strike, turning to the local community for support. They raised over $100,000 in strike support on GoFundMe in less than two weeks, a staggering amount given the size of the approximately two-hundred-patroller workforce.

The GoFundMe Park City Professional Ski Patrol Solidarity Fund was organized by patroller Brian Spieker. The GoFundMe page reads: "We remain in contract negotiations with Vail Resorts and we are committed to fighting until we secure fair and livable wages for our patrollers. The bargaining process has been ongoing for 16 months and we have seen our ski community show up again and again for us throughout this long process."[9] Ultimately, with strong community support, Park City union patrollers secured historic gains in their collective bargaining agreement with Vail Resorts, including adding more benefits and increasing base pay by almost 25 percent.[10]

In the early part of the 2022/23 season, lift maintenance workers at Park City Resort in Utah filed an election to join CWA Local 7781. The mechanics filed the petition for a representation election with the NLRB on October 11, 2022, with 80 percent of the unit signed on in support. In a press release, members cited chronic understaffing and low pay that is not commensurate with the precarity of their work.[11] The election was held in November 2022. Lift mechanics and electricians voted 35 to 6 in favor of unionizing—becoming the first ski resort lift maintenance department in the country to unionize.[12] "By unionizing, we are no longer passengers, but active participants in the direction Park City Mountain Resort and Vail Resorts Management Company are headed," Christopher Field, an intermediate electrician at Park City Mountain Resort, said. Liesl Jenkins, a lift mechanic at Park City Mountain Resort, added: "I'm incredibly excited to be able to advocate for ourselves as a department and enact real change through a contract."[13]

Patrollers have long hoped that the union movement that started in professional patrols would spread to other resort workers—something that is now happening as worker organizing spreads to lift mechanics and engineers. Workers in the whole industry need a collective voice—from ski instructors and lift maintenance workers to hospitality and service workers.

Union Members Begin Organizing to Protect Winter

When ski industry workers unionize, they get a voice on the job for a safer, more equitable workplace. But to protect their jobs and communities for the long term, unionized ski workers should use their collective voice to help build a movement for environmental sustainability.

In 2022, CWA Local 7781 members began organizing against climate change as a threat to their long-term job security. In the fall of that year, at a meeting on internal organizing in the run-up to the 2022/23 season, patrollers from Breckenridge and Park City discussed the need to build a political program to address big-picture issues like affordable housing and climate change. In November, union members overwhelmingly passed a resolution pledging to do their part to defend snowfall by fighting climate change.

To make this work possible, Local 7781 members are expanding their community alliances to include the environmental movement. This year, members have met with Conservation Colorado, the Colorado Sierra Club, and the Southwest Energy Efficiency Project, among others. CWA Local 7781 staff organizers are also getting mentorship and resources from the Green Workers Alliance, which mobilizes workers to build support for green jobs and policies.[14]

Fellow CWA members at the Breach Collective (CWA Local 7901)—a union cooperative that supports communities organizing on the front lines of the climate crisis—are also providing support. Breach Collective members are advising on strategy and communications for a campaign Local 7781 is leading to support local building electrification and affordable housing efforts. The campaign combines the environmental and economic concerns that directly affect Breckenridge patrollers.

Inside the larger union, ski industry organizers and Breach Collective workers are beginning to connect with union members in newly organized environmental nonprofits, such as the Sunrise Movement and the Audubon Society. In September, a cross-local group of workers met to network and connect about the environment. Their plan is to form an informal green network inside the union, which could eventually engage fellow union members around the struggle to protect the environment.

Similar work is happening in a number of different national unions, but we need to pick up the pace. Together, union workers can and must create space inside the labor movement for the honest conversation and urgent action their communities need on climate change.

Why Ski Industry Workers Matter to the Environmental Movement

The environmental movement needs workers to organize for the environment—not against it. Ski industry workers are strategically important because they are climate-vulnerable workers: they must organize against climate change to protect their jobs for the long term. Furthermore, ski industry workers challenge the fossil fuel industry narrative that jobs and climate are in opposition.

The labor movement is still deeply divided about the environment. Take, for example, the bitter berating of fellow unionists doled out by Laborers' International Union of North America (LIUNA) president Terry O'Sullivan in 2016 after several national unions (including CWA) came out against the Dakota Access Pipeline. O'Sullivan called those who opposed the pipeline "self-righteous unions," "bottom-feeding organizations," and "job-killing unions."[15]

The fossil fuel industry regularly uses its unionized workforce as a political shield. Take the January 2022 *Colorado Sun* op-ed entitled "Natural Gas Should Remain a Key Option in Colorado's Clean-Energy Mix."[16] The op-ed—which was jointly penned by fossil fuel industry and union leaders—sings the praises of natural gas.

Union workers with direct self-interest in saving the climate thus have a critical role to play in taking on their union siblings when they align themselves with fossil fuel corporate interests. And the environmental movement can help them do it. Together, climate-vulnerable workers and environmental organizers can counter the narrative that protecting jobs and protecting the environment are in tension. They can unite to engage the labor movement as a powerful force in the struggle against climate change.

In the labor-environmental movement ecosystem, ski industry workers are a bridge species that can straddle the boundaries of both movements. Other workers are similarly situated. New England fishermen, for example, are losing work because of climate fluctuations.[17] Western river rafters are negatively affected by drought.[18] Scuba diving operators are harmed by the destruction of coral reefs.[19]

The changing climate harms us all, of course, and all of labor must unite behind a just transition to a green economy that protects our world. But those directly impacted, like ski industry workers and others whose livelihoods depend on the great outdoors, can play an especially important role in fighting for an economy that does not destroy the natural world and imperil the future of their work. Union workers need a world with good jobs and a stable climate. We need more workers organizing for both.

14

THE FIGHT FOR AMERICA'S WORKERS MUST ALSO BE A FIGHT FOR ENVIRONMENTAL JUSTICE

Tefere Gebre

On April 29, 2022, I joined oil workers from the United Steelworkers Local 5 on the picket line to fight for a fair contract. As a longtime labor organizer, I had walked thousands of picket lines, but this was my first picket on the water. It was a classic Bay Area morning, with a chilly breeze off the bay and a bright sun to warm us. I climbed aboard a Greenpeace USA inflatable boat, ready to flank the five hundred workers who were on strike by confronting the oil tankers that had been making daily trips to the Chevron refinery in Richmond. As workers and community members, we found common ground in taking action to combat the exploitative greed of Chevron's executives.

The survival of our planet depends on our ability to bridge the fight for workers and the fight for a sustainable future. Too often, these fights feel unbridgeable. But that morning, I saw a new opening for how to connect our struggles. In the downtime between tankers approaching, when everything was quiet on the water, I asked the workers why they were striking. They told me the stories I'd heard countless times before: of health care costs skyrocketing, of paychecks not keeping up with inflation. But as the workers kept talking, they also brought up potential safety concerns at their workplace as well as their concerns about the pollution their work contributed to. As I listened to the workers, I thought about how corporations have pitted us against each other—kept us fighting over a slice of the pie when we could be struggling together for a larger and better pie. We all want the same things: a good job, clean air to breathe, safe and healthy communities.

In 2022, I joined Greenpeace USA as its chief program officer because, after decades in the labor movement, it had become clear to me that there will be no jobs left on a dead planet. In California, where I worked for many years as a labor leader, nearly thirteen million acres have burned in forest fires over the past ten years. That's one of every eight acres in the state—double what burned in California during the previous decade.[1] Meanwhile, record-breaking heat waves across the state have claimed thousands of lives.[2]

We are living in a climate emergency, and our only hope for survival is swift action across all levels of leadership to move our society off oil, gas, and coal. Unfortunately, I see no sign that fossil fuel CEOs want to make any such transition. Instead, they seem intent to drill every last drop, torch the planet, spend millions on lobbyists to influence our government, cash in golden parachute bonuses, and send taxpayers the bill for cleaning up their mess. Workers and community members need our plans and the movement infrastructure to deliver on them.

The transition from fossil fuels will hit workers first and worst; therefore workers and the labor movement must play a central role in bringing forward the solutions. Yet too often, voices in labor have been willing to talk only about the lowest-common-denominator climate policies, such as creating new union jobs in the green economy. I agree that we must rapidly increase the production of renewable energy and do so in a way that creates family-sustaining jobs in the green economy. But we must also be honest about the science: the worst impacts of the climate crisis are virtually guaranteed if we do not manage the decline of fossil fuel extraction and production. Once we accept that phasing out fossil fuels—through a planned reduction in fracking, drilling, and mining—is the bar we must meet, it becomes all the more necessary to work together to develop plans for investing in a just transition that protects the livelihoods of fossil fuel workers.

As an environmental and labor leader, I know we need both of our movements to win the kind of transformative changes we seek. But too often, we show up for each other in small moments—as I did when I joined the Chevron workers on the water that day—without building the infrastructure that can support an enduring climate-labor movement over the long haul.

Many environmentalists, for example, have demanded fair and equitable jobs for workers. But how many of us have built structures to ensure that workers are at the table defining what that transition looks like? How many of us have invested in sustained organizing to deliver material wins for workers? On the flip side, many labor organizers have been too timid about spelling out what it will really take to forestall the worst effects of climate change. As a result, too few labor organizations have invested resources in fighting for a just transition—focusing instead on protecting the little that we have on a dying planet.

But for all of us to survive—and thrive—labor and climate leaders need to develop a movement-building culture that is centered around shared goals and sustained alignment beyond one constituency's interests. We will need to bring many different people with different interests together to develop lasting relationships that can sustain our movement. Doing so requires us to move beyond tactical, short-term alliances to develop the infrastructure that supports labor-environmental coalitions through multiple campaigns and election cycles.

Building Lasting Coalitions

Too often, the labor and climate movements aren't organizing together because we think it will be too hard or that there's too much that divides us. But I've been working on hard and improbable campaigns for much of my career. What I've learned is that we can organize anywhere in the country to win power for working people if we focus on building a movement of community members and workers.

In 2006, I quit my job as the Southern California director of the California Labor Federation to become the political director of the Orange County Labor Council. Orange County is a birthplace of the modern conservative movement. But it's also home to a growing immigrant and underclass community. For every one of the OC housewives you see on TV, there is someone driving their kids to school, cleaning the pool, cooking in the kitchen, or racing to stock shelves at the grocery store. I set out to organize those workers, a vast majority of them immigrants like me.

When I arrived, the Labor Council was largely dormant and underperforming. I knew we needed a longer-term approach to building worker power than just collecting cards for individual union elections. So I started looking for what I call *vacuums*—the gaps in our long-term movement-building infrastructure. For example, when I first started, the Labor Council office was in a small outer building of another local union—on an island by ourselves. I convinced another union to go in on a new building in the center of town that was furnished with phones and lots of rooms. We called up local community groups one by one to ask if they needed an office or meeting space. We could see that our movement was fractured, and we wanted to be a hub of progressive organizing.

We also began a new program organizing with the significant communities of faith in the region, whom the Labor Council hadn't previously engaged with. So I hired two clergy organizers to build relationships in communities of faith. To bolster local engagement with religious leaders, we created a new chapter of Clergy and Laity United for Economic Justice (CLUEJ) that was housed at the Labor Council. We also brought a strategic political lens to our faith-based organizing: we mapped out where every legislator in the county worshiped and enlisted organizers and community members to organize them in their places of worship.

As we expanded our community-labor organizing, we needed a locally focused organization to hold these emerging partnerships together. We started Orange County Communities Organized for Responsible Development (OCCORD) to bring the Labor Council, unions, environmentalists, and community organizations together to pressure local governments to ensure economic development plans would benefit the community.

Within the Labor Council, we also began a long-term political outreach program called the Voter Infrastructure Project (VIP), a team of permanent canvassers who did precinct walking six hours a day every day. Every morning, the canvassers would pick up a new precinct map, spend all day registering voters, and ask them eight survey questions. When they came back to the Labor Council, they spent two hours writing personalized postcards to every person who opened the door. In six years, we registered 140,000 voters.

We didn't just talk to the voters, we also engaged elected officials. We required anyone seeking our endorsement to take a six-hour course on workers' issues. Since this was a pre-endorsement requirement, half of the people who went through the class didn't even get endorsed. But they still got the education.

This kind of long-term power building provided the crucial building blocks for one of the most righteous fights I've ever been a part of. In 2011, I visited one of the worksites of sanitation workers, who included trash truck drivers and landfill workers. Our site guide explained that back in 1999, a local union had tried to organize these workers. They had collected the number of cards necessary to force an NLRB election, but two weeks before the election was set to take place, the workers believed that management called Immigration and Customs Enforcement (ICE). The worksite was raided, and almost all of the workers were deported. Devastated, the union had not tried to organize the workers since.

The working conditions were brutal and humiliating. The workers spent eight hours a day outside, in the Southern California sun, searching through trash bins for recyclable items from regular residential and commercial trash. The workers were issued one pair of gloves, which generally didn't last more than two hours, and one bottle of water per eight-hour shift. But one of the worst and most degrading conditions of their jobs was that, every day, each person had to pick a time to go to the bathroom. They would have to write that time on the board, and they could not deviate from the set time. I kept thinking, *If we can't fight for these workers, what is the point of the labor movement?*

My communications director convinced a few of the workers to share their struggles with us, and we quickly organized a meeting of our community and clergy leaders, who became determined to help them organize. The coalition sent a delegation to demand management sign an agreement not to use ICE to retaliate against the workers. Because of how much community and worker power we had built in Orange County, management agreed.

With the agreement not to retaliate in hand, we started to help the workers organize. Seven months after I first met the workers, 78 percent of them voted for the union. The workers negotiated their first contract at the Labor Federation office, surrounded by community members. That first contract provided the workers with health care for them and their families, two weeks of paid vacation, paid sick leave, more than double their previous take-home pay, a pension, and the dignity of knowing they had a democracy at their workplace. And the bathroom board—a symbol of the unjust, unfair, and degrading treatment of the workers—was gone.

A few lessons stand out from that campaign for those of us seeking to build a labor-climate movement. First, we multiplied our people power by bringing existing organizations together in a movement united by shared goals and plans. Second, we pushed ahead on our community partnerships, even when it alienated allies. When I started pulling together clergy and community groups into the Labor Council, two large union affiliates left because they thought we needed to organize workers, not communities. After we won the campaign for sanitation workers—along with other campaigns focused on community issues—those two unions rejoined, and a couple of other unions affiliated with the Labor Council for the first time. Sometimes coalition building looks like a risk, but it builds our power for the long term, and allies will come around once they see that power in action. Third, we focused on building a movement, including infrastructure for growing our political power. When we wanted to tackle a campaign around workers' issues—or racial or social justice—we were ready to leverage that power.

By doing so, we became the leading labor council in the state of California for voter outreach and membership growth for five years in a row. Most importantly, everyone benefited. Together, workers and the community won a ban on big-box stores, something many people thought was impossible in Orange County. After a long struggle, the community replaced the entire Costa Mesa City Council, which had a history of pushing apparently racist policies, like trying to restrict taco trucks from the city limits. Orange County eventually started electing progressives to county, state, and federal offices. The tide turned.

Unions around the country—like United Teachers Los Angeles (UTLA) and Service Employees International Union (SEIU) Local 26—are already building these kinds of labor-community partnerships to win on climate justice, as you'll read in other contributions to this anthology. This work needs to happen in many more places for us to build a movement at the scale necessary to address the climate crisis.

Shared Goals for a Pro-Worker, Pro-Climate Economy

As we work to build broad, cross-sectoral movement power, we need shared goals to unite around. What is our pro-worker climate agenda? Of course, the real answers to that question will be decided by the people on the ground—by union and community members and leaders at the local tables who have a strategic view of the solutions most relevant to their communities.

But too many times, I have heard environmentalists default to an agenda for a just transition that lacks breadth, specificity, and boldness. For example, new solar panel installation jobs are not sufficient when the workers make only a fraction of the pay that they would at a refinery. While I certainly don't speak for every worker, I have walked many precincts and listened to thousands of workers and labor leaders in communities whose local economies are or historically have been dependent on fossil fuel production. From that intent listening, I would like to propose a handful of solutions that, if implemented properly, with good stakeholder engagement along the way, might meet many of the needs of workers and communities impacted by the transition away from fossil fuel production.

First, as others have said, we must deliver bold investments in truly sustainable industries, coupled with strong labor standards. There is so much potential for good-paying union job creation in the electrification of our ports, buildings, homes, and schools. If targeted to historically energy-producing regions, some of the same investments that are needed to combat the climate crisis could become a lifeline to diversify and revitalize those same local economies. Incoming support from the federal government (from both the Infrastructure Investment and Jobs Act and the Inflation Reduction Act) could create hundreds of thousands of union jobs at the forefront of the clean energy revolution and across sectors such as manufacturing, transportation, land remediation, and infrastructure cleanup, and other potentially sustainable industries such as offshore wind development. It is urgent that these funds be deployed quickly and with all available labor standards attached to ensure that good union jobs are ready and waiting for any worker who is affected by the contraction of the fossil fuel industry.

If we use all the available policy opportunities to win project labor agreements, community benefit agreements, and community workforce agreements and to expand apprenticeship programs, we can create new, family-sustaining jobs while building high-quality green infrastructure. Such projects have the potential to bring together community members, workers, and businesses around common interests: cleaner air, healthier neighborhoods, booming business, and stable, long-term careers for workers.

Second, we must deliver funds to workers and communities proactively and directly. A disorganized transition is going to disproportionately affect workers in existing industries. These workers and communities have done the hard work of powering our economy for generations. They must be honored, respected, and supported during the transition away from fossil fuels. This includes support for people in all stages of their careers, from training for early-career workers to pension guarantees for those approaching retirement. How about fully funded pensions, wage replacement guarantees to make up for salary gaps, paid health care benefits, priority hire for new jobs, free college and trade school, and relocation expense coverage? How about a fund to replace lost tax revenue for communities that have relied on the fossil fuel industry to support critical public services?

Some unions are already leading in defining a potential blueprint for worker and community support through the transition. In June 2021, twenty California unions—including unions that represent fossil fuel workers, teachers, nurses, janitors, and home care workers—unveiled their California Climate Jobs Plan for how California can, over the next decade, create over four hundred thousand jobs per year while cutting climate emissions in half. Alongside that plan, the coalition outlined how an equitable transition program could provide effective relief for displaced fossil fuel workers as well as their families and neighbors.[3] Their research showed that a plan that takes into account voluntary retirements will be cheaper and more worker-friendly than the episodic or abrupt closures that are virtually inevitable if we leave the transition to the vulture CEOs. Another inspiring example is the work of the Colorado AFL-CIO, which, alongside other unions and environmental and community groups, helped to pass a law to create the Office of Just Transition in the Colorado government to assist communities and workers affected by the decline of the coal industry.

Putting Workers at the Center

At Greenpeace, we have a lot to figure out about how to center workers in our campaigns and to build long-term partnerships across the labor and environmental movements.

One concept we are beginning to explore is the creation of a workers' council—a structure for workers and labor leaders to weigh in on the work of the environmental movement. The Big

Greens don't always have rank-and-file members who participate in the way that union members do, so we need a formalized way for workers to advise us on the policies and programs we're pursuing. The workers' council is a work in progress, but the idea is that it will offer workers a way to vet the environmental movement's plans and create a platform for ongoing dialogue beyond the staff of environmental and labor organizations. I'm convinced that we will leave this process with valuable feedback and a more finely tuned political compass. But what's more, we might discover that we share a lot of common ground in our vision for thriving, regenerative economies.

For too long, the corporate leaders of the fossil fuel industry have pretended to care about workers' interests, all to prop up their own bottom line. The environmental movement must disrupt that dynamic by showing that we are serious about bringing workers to the table to build the power we need to deliver material benefits on a timeline that will matter.

When I joined the labor movement three decades ago, there weren't many folks who looked like me—a darker-skinned immigrant. The reason I got involved was to help shape a future that takes bold action to confront injustice. It's time to make sure no one is forgotten or left behind as we make our dream of a green and peaceful planet a reality.

We have a responsibility to listen to and partner with immigrants and communities of color to create new and fresh approaches to the challenges we face today. I have no illusions that phasing out fossil fuels in the United States or anywhere in the world will be politically easy; however, the dueling emergencies of the climate crisis and the public health crisis caused by drilling for fossil fuels must be addressed with bold actions before it is too late. There is no way to prosper on a dead planet. The work ahead is to figure out how to build alliances—from picket lines to state houses—to push for something the fossil fuel industry will never voluntarily do: create a timeline, dedicated funding, and a plan to phase in new union jobs as we phase out fossil fuels.

Benjamin Smith, senior campaigner for strategic partnerships at Greenpeace USA, contributed to this article.

15

DON'T WASTE LA

AN INTERVIEW WITH LAUREN AHKIAM

Miya Yoshitani

Lauren Ahkiam is campaign director of Water Justice LA at the Los Angeles Alliance for a New Economy (LAANE), an organization fighting for economic, racial, and environmental justice. She started working at LAANE in 2012 as a researcher on the waste and recycling team.

Can you give us a little bit of history on LAANE?

We were founded thirty years ago in the wake of the racial justice uprisings in Los Angeles. LAANE was co-founded by Maria Elena Durazo, who at the time was leading the hospitality workers' union and the Los Angeles County Federation of Labor, along with Madeline Janis and others. She [Durazo] was really looking at the power of labor coming together with community because union members are community members, community members need access to good union jobs. LAANE later expanded to take on environmental justice campaigns. There is so much opportunity for frontline communities, mainstream environmental groups, and the labor unions representing workers in those industries to collectively confront corporations. A lot of our work focuses on specific industries, using a model based on deep research, coalition organizing, strategic communications, and policy advocacy.

LAANE pulled together the Don't Waste LA coalition. Can you give us some of the background on that campaign?

The first environmental campaign that LAANE engaged in was at the port. We worked with the Natural Resources Defense Council (NRDC), the Teamsters, East Yard Communities for Environmental Justice, and other groups to confront the impacts of port trucking on neighborhood health and driver health from diesel trucks, as well as the rampant misclassification of workers as independent contractors by those port trucking companies. NRDC and Teamsters Local 396 started to look at the waste industry as another place where there's a lot of environmental justice impact and a lot of worker exploitation—a place where the city has a role it can play in how that industry is regulated.

LAANE started with research and community organizing, like talking to partners like Pacoima Beautiful—where I worked at the time—about the neighborhood impacts of the waste and recycling industry. We were really concerned about the amount of diesel traffic in the community. The sorting facilities are also, along with landfills, predominantly located in communities of color and working-class communities.

LAANE pulled together the Don't Waste LA coalition with groups like Pacoima Beautiful, NRDC, and the Teamsters to comprehensively affect the industry at the larger geographic scale. What we discovered is that there was widespread finger pointing within the industry because the structure of the city's commercial waste system was very much a race-to-the-bottom, unregulated free-for-all. There were dozens of different companies that were operating; some of them were highly capitalized national companies, but others were pretty small family operations with two trucks. We would hear anecdotes about them dumping trash to save money, not paying the tipping fees at a landfill to have it properly disposed of, not properly paying drivers, and expecting drivers to work really wild hours that are very unsafe. There was the hiring of day laborers to work in the sorting facilities and not providing proper protective equipment. We would see ten or twelve different companies collecting trash on a given block in any given week. Instead of one truck coming once a week, you would have twelve different trucks coming all different days. So the air quality impacts as well as the road impacts and the traffic impacts were really illogical.

What was the demand that you came up with?
San Francisco has an exclusive franchise contract system, so that was the high standard based on our research. We would talk about it in terms of good jobs, clean air, and recycling for all. And we were really excited to work on it because it has so many different impacts, from food security to traffic on the street.

What were some of the benefits to workers of winning this policy change?
Dozens of very low-road operators could no longer get work in the city of LA. You don't really want a mom-and-pop waste hauler. You want a pretty sizable institution, well capitalized. And those well-capitalized, sizable waste haulers are more likely to be union. It's also fewer employers for the union to work on organizing. So some companies that were not unionized, as soon as they got the contracts, they went union, because there was greater certainty of how much work they would have for the next twenty years, and there was a more level playing field for worker compensation and benefits with less fear of being undercut by a low-road bidder.

And then if haulers are in violation of different pieces of their contract with the city, that could trigger them losing their contract. There's just a lot more accountability.

Another thing that is a requirement in the new system is to not have what's called Dirty MRFing, which is a materials recovery facility that is separating recycling and trash. Low-paid women were working on those sort lines trying to grab out the recyclables and trying to salvage aprons out of the waste stream to protect them, because they're not given uniforms. If their earplugs fall out—because it's super loud—the boss would say, "You can't get new earplugs until next week. We are not going to just give you new earplugs all the time." People would get pricked by needles and then be worried for weeks of what that could mean for their health.

As well as exposure to toxins dumped right into garbage cans and dumpsters . . .

Exactly, like a bag gets ripped open and it's full of asbestos or ceiling tiles or something.

There was another story of an organics facility that was in the Central Valley but was processing the organics waste from the city of LA. They sent brothers into a ventilation line to check something out. And there was so much toxic gas from the improperly managed waste that they both died, and they were not even old enough to work there. They were like sixteen and eighteen. It was really heartbreaking.

The waste industry has some really high fatality rates. I think at the time it had the third- or fourth-highest fatality rate because of malfunctioning trucks rolling back on people or people getting fingers cut off in the sorting facilities. And in communities like Pacoima, the people living there are also working as drivers or in those facilities because that's some of the local jobs and so they're being doubly impacted or triply impacted.

When we do this kind of organizing that gets to the heart of what frontline communities and workers really need, we get the one-off wins, but we are also able to build stronger unions, stronger organizations that can take on the next win and have more transformational wins in the future.
Yeah, absolutely. One of the things I think of first is just building relationships between community groups, environmental organizations, and labor organizations, and having it where there's so much alignment. Frankly, it's very rare how much alignment there was on this campaign. The more you recycle, the more people it takes to pick up the recycling. And the more regulated it is, the better for everyone, so it's very mutual.

Are there any big opportunities on the horizon that you see for replicating that combination of interests between workers and frontline communities around climate solutions?
The work I've been doing more recently on water is very similar. We need to invest in our water system to be more climate resilient after decades of disinvestment by the federal government or the state being cash strapped because of bad tax laws. There's an opportunity for union workers to build things like large-scale stormwater capture projects. Then there are the investments we need in climate-resilient buildings to make our public institutions, like our schools, more efficient—and to do that in ways that create union jobs. There are so many different building trades that have expertise in how our buildings work and what we need to do to get them up to snuff.

Is there anything else that you want to share?
This is a long journey we're on. We need to do the work and the work is really hard, but it helps when we are also in a loving relationship with each other. And I think that this campaign is just a really nice example of that, of persevering in the face of a lot of very uphill stuff in a way that forged really tight bonds.

16

BUILDING A WORKER-LED MOVEMENT FOR CLIMATE JUSTICE

Brooke Anderson

I grew up in East Central Illinois, several hours' drive from Chicago through monocropped corn and soybean fields. At seventeen years old, I got involved in an environmental justice fight to shut down two medical waste incinerators in my hometown. When incinerators burn plastic—like IV bags or medical tubing—they release dioxin into the air. The dioxin was wreaking havoc on the health of the mostly working-class Black neighborhoods surrounding the two not-for-profit hospitals. The fight was led by two community organizations: Champaign County Health Care Consumers and United Citizens and Neighbors. Most of those involved were parents trying to protect their kids from headaches, nosebleeds, and cancer—all of which were on the rise. They taught me to organize: to film the plumes of smoke coming from the incinerators, go door-to-door with health surveys, hold community town hall meetings, and organize marches. Eventually, we shut down the incinerators.

I knew that whatever we'd just done—building grassroots power—was what I wanted to do for life. I also needed to pay rent, so I got a job as a union organizer. I worked for two unions in

Illinois before moving to Oakland in 2005. Over the course of the next decade, I worked with mostly low-wage, immigrant hospitality and port workers with the East Bay Alliance for Sustainable Economy (EBASE), in partnership with UNITE HERE 2850 and Change to Win/Teamsters. What I quickly saw was that as the hospitality and transportation industries grappled with the consequences of ecological disruption, they moved to shift the burden on to the low-wage workers who had least contributed to the problem and who had the least resources to shoulder the weight of the transition.

My journey as a young environmental justice activist turned union organizer reflects the question many in labor are grappling with: What does it mean to fight for dignity in a workplace on a rapidly deteriorating planet?

Two examples:

In 2006, I was organizing hotel housekeepers on a campaign led by EBASE, together with UNITE HERE 2850. We passed a living wage in the city of Emeryville, then engaged in a multiyear fight to implement it. One hotel fired our entire organizing committee ten days before Christmas and sent federal Immigration and Customs Enforcement (ICE) agents to workers' homes to scare them. In the midst of these fights, hotels across the country started implementing the "Green Program," which encourages guests to elect not to wash their sheets and towels in order to save water, often incentivized with a coupon for a free coffee in the hotel's cafe. With less laundry to do, many hotels then either laid off laundry workers or significantly cut their hours. Meanwhile, if you asked workers about the sustainability of the hotel industry, they would report that they were overworked, that their bodies were exhausted by their thirties, that they cleaned with toxic chemicals, that the hotels engaged in massive food waste, and that recycling was routinely dumped into the trash. The hotels cut labor costs under the greenwashed guise of sustainability while consolidating wealth and power.

Four years later, I worked as an EBASE organizer on a Change to Win/Teamsters campaign to organize port truck drivers at the nation's five largest ports. In Oakland, two thousand truck drivers were driving old beat-up diesel trucks that spewed emissions through the predominantly working-class Black and Brown, already over-polluted neighborhoods of West Oakland—causing high rates of asthma, cancer, heart disease, and respiratory issues. The state passed desperately needed clean air regulations to upgrade to a new green fleet of trucks. But because the drivers were misclassified as independent contractors, the burden of buying those new green trucks fell to them. Many drivers—paid by the load, not by the hour—made $20,000 per year. New trucks cost $100,000. Drivers, together with the Teamsters, fought to be reclassified as employees in advance of the regulations but were unsuccessful. When the regulations went into effect on January 1, 2010, an estimated eight hundred drivers lost their jobs in one day. The burden of transitioning those trucks was put on drivers rather than on Walmart, Target, Home Depot, and other big multinational corporations whose goods they hauled.

These examples are indicative of trends I was seeing across the labor movement. Workers who had contributed the least to ecological erosion and had the least resources to shoulder a just transition were being asked to sacrifice the most for more sustainable industries. Meanwhile, corporations were amassing wealth and power under the guise of sustainability. This wasn't a just transition; it was green capitalism.

For five years, a group of rank-and-file workers and union organizers in the Bay Area built a worker-led movement for a just transition. Our organization, called Climate Workers, was a project of Movement Generation Justice & Ecology Project. Rather than trying to make sure "green jobs" went union, as many in the labor movement had done, or trying to get unions to pass climate solidarity resolutions, as many environmental organizers had done, we created a structure for union members to fight together on climate.

We shut down the project in 2019, but the work we set out to do—organizing workers to fight for climate justice—remains more urgent than ever. There are three key lessons I learned as the project director of Climate Workers about how the labor movement can organize to transition from an economy built on extraction to one grounded in ecology, a regenerative relationship with land, and self-governance of our labor.

1. **Fight the bad and build the new.** Labor must go beyond making sure new "green" work goes union. We must mobilize at scale to transition out of extractive industries ("fight the bad") and lead with a vision of how labor can heal people and the planet ("build the new").

2. **Take positions—and pick fights—outside of what labor leadership deems politically possible. Then win, shifting what is politically possible.** To do this, we must organize not just at the level of locals, central labor councils, state federations, and international unions. We must also build a strong, multi-union, rank-and-file leadership body committed to climate justice that is capable of playing outside the bounds of what is currently deemed "politically possible." We must then win and bring our unions along with us.

3. **Talk directly about the history—and current contours—of white supremacy in the labor movement.** We cannot advance a just transition in labor without reckoning with the role of race and racism. While there are certainly nuances by geography and industry, many living-wage jobs in fossil fuel exploration, extraction, production, and distribution go to white workers. Those in the toxic path and sacrifice zones of the fossil fuel industry are overwhelmingly Black, immigrant, and Indigenous workers who are often historically excluded from unions or are unionized but still making less than what it takes to thrive in their communities. When we prioritize the preservation or creation of jobs in ecologically devastating industries over the health of communities impacted by those industries, we often exacerbate existing racial and economic inequities in the labor movement.

Even though most of the labor movement represents workers whose lives have been negatively affected by extractive industries, the unspoken agreement in the labor movement has long been that it is only those unions whose members' jobs are at stake that can speak for labor on a given issue. What that means is that the concerns of the Building Trades (whose members are predominantly highly paid white men, with some regional variation) have taken precedence over the concerns of workers who make lower wages and are more likely to be women, immigrants, Indigenous people, or other people of color.

A New Model for Climate Action

It is only through the exploitation and coercion of human labor that the extractive economy has been able to concentrate wealth and power to unleash massive destruction on ecosystems worldwide. As such, any just transition must reclaim our labor—literally, to free it from the chains of the market and apply it to meeting the needs of our communities and restoring our ecosystems. Simply put, we must return our labor to the web of life. Climate Workers emerged out of this vision for a just transition.

At the time we launched Climate Workers, most of the environmental action in the labor movement had been in the model of the BlueGreen Alliance (BGA), an organization that seeks to facilitate dialogue between unions and environmental groups. What the BGA model succeeds at is making sure that climate jobs go union. But BGA and other groups in this model have struggled to cohere the labor movement around either a more comprehensive strategy for a just transition or a position against extractive projects like the Keystone XL and Dakota Access pipelines, fracking, and mountaintop removal. One of the limitations of BGA is its coalition-based structure, in which big unions—like the United Steelworkers (USW) and the Service Employees International Union (SEIU)—act as members. Some of those unions have members employed in extractive industries, which makes it difficult for the unions to take a position against the expansion of those industries.

Climate Workers experimented with a different model. First, Climate Workers was local in scope. We aimed to be—but never became—a replicable, translocal organizing model. Second, our members were individual rank-and-file workers and union organizers, not whole unions or organizations. Correspondingly, we looked to do slow, patient political education and base building to develop the climate justice muscle of unions from the bottom up rather than to take shortcuts to get key labor leaders on board. Third, we intentionally focused on moving the unions whose members were most impacted by the extractive economy, not the unions whose members held jobs in those industries. As a consequence—or maybe as a starting place—we put race, class, and gender squarely in the center of the debate about jobs and climate. And finally, we looked for ways in which workers could reclaim their labor from the extractive economy and experience what it looked and felt like to have that labor serve the broader community and be aligned with the needs and realities of the living world, rather than the interest of profit.

Here are some examples of the work we did:

- Hundreds of rank-and-file union members marched with us against the Keystone XL pipeline and the expansion of the Chevron refinery in Richmond in 2014, for our first action.
- Farmworkers, day laborers, landscapers, domestic workers, fast-food workers, city workers, librarians, wastewater treatment plant workers, and others lent their voices and stories to our *Worker Wisdom in a Changing Climate* series, featuring original interviews with union members and other workers on the front lines of climate change.
- Union workers at Casino San Pablo took over an abandoned lot in Richmond, planted fruit trees and herb spirals, and created a "strike garden."
- Fast-food workers and urban farmers shut down a McDonald's drive-through, handing out hundreds of free, locally grown organic burritos.
- Labor leaders caravanned down to the fracking fields of the Central Valley and later marched ten thousand strong

in the streets of Oakland, blockading the doors of the state building to call for an end to fracking.

- Walmart workers went on strike on Black Friday, backed by hundreds of environmental justice allies in Richmond, calling out Walmart as the world's largest climate criminal.
- We fought for and won a significant expansion of Oakland's citywide compost program—from single-family homes to multifamily units. In addition, we raised the wage for Oakland workers, won over $1 million in back wages for previously temporary recycling workers, and won grants that shifted compost education and outreach jobs to immigrant women recycling workers.
- Refinery workers went on strike at the Tesoro refinery in Martinez; hundreds of environmental justice leaders joined them, backing their demands around safety and public health.
- Hundreds of workers shut down Wells Fargo and Bank of America in downtown Oakland in solidarity with Water Protectors fighting the Dakota Access pipeline.
- Workers, residents, and youth stopped Phil Tagami's plan to bring dirty coal through Oakland's working-class Black and Brown neighborhoods.
- Thousands of Bay Area workers from across dozens of industries, backed by nearly one hundred environmental and climate justice organizations, went on strike on May Day in 2017, in response to the Trump administration's growing attacks on immigrants. Climate Workers organized a press conference calling on corporations not to retaliate against striking workers.
- Dozens of unions brought us in to train their members on climate justice, and dozens of worker leaders and

union organizers went through our intensive five-day just transition retreats. We conducted extensive political education for union and non-union workers about the scope, scale, urgency, and pace of the unfolding ecological crisis and its roots in the exploitation of human labor.
- We educated everyone from big environmental organizations to local environmental justice groups to national climate justice formations about how to work with labor. For example, we wrote a tip sheet for environmental groups on how to work with labor, and we ran a training based on that handout for groups like the Sierra Club, 350.org, and the Climate Justice Alliance.

There were four main components to our work: political education and base building, fighting the bad, building the new, and changing the narrative. These categories were not necessarily distinct. For instance, political education and base building were at the core of everything we did. But it's helpful to name these categories as shorthand for talking about what worked and what didn't—where we had traction or momentum and where we struggled.

Here are a few of my personal reflections on what worked and what didn't and the lessons we can draw from our experience.

What Worked

Climate Workers significantly moved labor on key "fight the bad" campaigns, like opposing the Keystone XL and Dakota Access pipelines, fracking, and No Coal in Oakland. These campaigns, though mostly waged locally, created space for locals elsewhere to take progressive stands and set in motion a cohort of organizers committed to a just transition, many of whom are now in leadership positions throughout the labor movement. The lesson is that you don't have to organize everything and everyone. You just have to move the landscape enough to show cracks in labor's unanimity. Once there is room for dialogue about "what labor's position should be"—as opposed to "labor's position is X"—we found that what is right usually prevailed, particularly if we talked about race, class, and gender. This was especially true in the campaign to defeat the proposed coal terminal in Oakland. When the terminal was first proposed, many in local labor deferred to the building trades, which supported the project because it would mean jobs for their members. However, once unions with members in the path of the coal trains—often lower-wage, Black, and Brown members—spoke out against the project, it put labor's position back up for grabs. We eventually moved much of labor to oppose the terminal, and the city of Oakland followed suit, rejecting the proposal.

These "fight the bad" campaigns allowed us to incorporate workers and union members into campaigns that no single union had the capacity to take on alone. In other words, on many issues, we had only five union members from each of fifteen unions, as opposed to needing one hundred members from one union. The advantage of this individual membership model (that is, where individual workers, not unions, are members) is that it created a political home for workers and organizers whose unions were not yet "there" on climate to connect, strategize, and move political work together. In doing so, they brought their unions along. Our model also allowed us to take stands on issues that we would never have been able to take if we were waiting on coalition partners. To name just one example, one of the most supportive unions during our fight against fracking could not take a public position because they feared angering the governor at a key point in their own political battles. As a result, they sent all of their members to participate as Climate Workers members. The downside of the individual membership model was that we often attracted joiners, not leaders. Our membership was never truly representative of our base, nor did we have wide and deep organizational buy-in except from a very few, often already progressive, unions.

At times, the official structures of labor worried that we were getting out ahead of labor's stated position on climate issues. By creating a "left flank" on climate within labor, we got the more moderate arm of the labor movement worried that they looked behind on climate. This often moved them to take positions that they would not have taken otherwise. On the flip side, Climate Workers—and especially our narrative series *Worker Wisdom*—served as a mirror to the rest of the labor movement. In essence, our work said: *Your members really do care about this, they are coming out to our events, and they are hungry for labor-led action for climate justice. Join us!*

We also created experiences for workers to put their own labor to use to meet their needs and the needs of their communities. For example, workers took over an abandoned lot and turned it into a strike garden to grow their own food during a strike at the Casino San Pablo. As another example, during the fast-food worker strike, striking workers prepared three hundred burritos from food grown at the Gill Tract Farm in Albany, applying their labor toward building the world they wanted to live in—with community gardens, cooperative labor, and healthy food for the hood.

What We Struggled With

While we excelled at bringing unions in to "fight the bad" campaigns, doing so allowed them to dip into climate justice without committing to the harder work of "building the new": industry-level, comprehensive campaigns to radically redesign the industries in which they represented workers (such as hospitality, transit, health care, education, and security). We wanted to implement a just transition campaign with labor at the front, leading with a vision of how to reclaim the skills of their members from the extractive economy and put them in service of the community and ecological restoration. Running that kind of campaign takes a deep commitment on the part of the entire union, but especially the leadership, to drive a multiyear, resourced campaign. In other words, Climate Workers and the union would have needed to identify what the new organizing win would be through a just transition campaign (such as organizing upward of one thousand workers) and be able to offer significant staff capacity to move it. It was challenging for unions to make that commitment for several reasons:

1. **External threats:** Unions were facing down *Janus* (a Supreme Court decision making it harder for public sector unions to organize)—an existential crisis of epic proportions that led many unions to shrink in budget as well as in sense of purpose. That is, they returned to a focus on bread-and-butter issues and membership drives.

2. **Internal threats:** Union leadership is not like nonprofit leadership. Leaders are elected, not hired (which is a good thing!). In many of the unions we were working with, leadership was acutely aware that they would soon be up for reelection and that their members wanted them to win concrete gains in wages, benefits, and working conditions. Members sometimes viewed the union's work on other social issues (such as climate)

as a distraction or a waste of dues. With political education, workers often became deeply invested in and loyal to climate justice work, but it took consistent, systematic engagement. In some of these unions, election campaigns came amidst deep divisions, both personal and ideological. One of the union leaders we worked closely with for years was voted out of office specifically for his work driving a just transition campaign in his local (even though the campaign brought the union more members).

3. **Pressure from their internationals:** Many of the unions we worked with faced tremendous pressure from their internationals not to take a stand or invest resources on campaigns that might threaten the jobs of union members elsewhere in the union or that might upset allied internationals. One international even fired some of its local organizing staff for having worked too closely with Climate Workers and with other local organizations on issues like gentrification and Black Lives Matter.

4. **The intensity of crisis in workers' lives:** Rank-and-file workers—especially those most affected by both economic and ecological crises—are constantly inundated with everything that makes life hard: living far away from work, working multiple jobs, raising kids as single parents, caring for sick elders, facing constant threats of eviction or deportation. The realities of working poverty make it hard for workers to participate in their union on the most bread-and-butter issues, let alone on issues that feel one step removed.

Reflections and Questions

In the few years that Climate Workers existed, we learned that it is valuable just to try something—to change it up, experiment, and see what works. Though Climate Workers has since closed, many of its organizing committee members took the politics and strategies we developed into other organizing spaces, most notably the North Bay Jobs with Justice's work with Indigenous and immigrant farmworkers in nearby Sonoma County, California.

More than anything else, we learned that workers get it. They get that ecological erosion is upon us in epic proportions and that reclaiming our labor is at the heart of any project for restoration. They get it because of who they are and where they are positioned in the world. They know that capitalism isn't working for them, that corporations take more from the land and from workers faster than either can regenerate, and that workers come from countries, cultures, and histories tied closer to the land, with more direct knowledge of the devastation wreaked by multinational corporations, colonization, and free trade agreements.

Paid union organizers are also terrified of the ecological crisis, but many doubt that workers will vibe with climate justice. Others are just too overwhelmed with the basic work of the union to add another thing to their plate. Climate Workers taking on responsibility for political education of rank-and-file workers—and the creation of a political home for workers to engage in climate justice—was critical. Yet that alone wasn't enough to convince labor to lead an industry-level, multiyear just transition campaign.

The models that labor currently relies on are insufficient to catalyze the type and scale of change that we need. We trust that there will be hundreds of different attempts to build a just transition in labor. And yet, the question remains: Can organized labor—as it exists today, with all its challenges and constraints—be a home for a just transition at scale? If so, how do we get to scale? If not, where else should we be building?

17

LISTENING TO THE LAND, LISTENING TO THE WORKERS

FARMWORKER ORGANIZING ON THE FRONT LINES OF CLIMATE CRISIS

Davida Sotelo Escobedo, Max Bell Alper, Davin Cárdenas, and Aura Aguilar

Anayeli Guzman was working the wine grape harvest in Sonoma County when the Kincade Fire came roaring down the parched slopes of the Mayacamas Mountains. Fueled by hurricane-force winds, the fire burned for two weeks in late October 2019, forcing the evacuation of two hundred thousand residents. But wine growers sent farmworkers like Anayeli back into evacuation zones to ensure they could make the harvest. Anayeli recalls her experience working late at night in thick smoke.

> El humo te arde la garganta. Te arde los pulmones cuando tienes que respirar. Y nadie se preocupa. Simplemente nos piden que hagamos el pedido que necesitan. (The smoke burns your throat. It burns your lungs when you have to breathe. And no one cares. They just tell us to get the job done.)[1]

When Anayeli got a text from her babysitter saying she was evacuating with Anayeli's daughter, she faced a decision familiar to all farmworkers: go to safety with her family and risk losing her job or stay behind to work in dangerous conditions.

Anayeli's experience is all too common. The skilled labor of over eleven thousand immigrant and Indigenous farmworkers supports the region's multibillion-dollar wine industry—often in hazardous conditions. Sonoma's grape harvest season—between early August and early November—lines up with the most dangerous period for wildfires in the region, putting workers' health at risk and amplifying the poor working conditions they already experience at the hands of abusive bosses.

In the spring of 2021, a core team of seven farmworker leaders with North Bay Jobs with Justice began building a base to push for improved working conditions during wildfires. They called their demands the 5 for Farmworkers: language justice, disaster insurance, community safety observers, hazard pay, and clean water and bathrooms. After a year of organizing, organizers and worker-leaders brought over one thousand farmworkers from multiple employers into the campaign network; over three hundred farmworkers participated in direct public actions to push for the demands. By the end of the 2022 harvest—and despite attempts by the industry to discredit the campaign—workers won historic victories across all five demands.

As farmworkers gained momentum around immediate victories, we began to think about the long-term strategy for workers to build resilience in the face of escalating climate crises. On top of wildfires, persistent drought and extreme weather impact harvest yields, making it harder for many workers to find steady employment. The wine industry has long been a culprit of extractive agricultural practices—overdrawing from diminishing water supplies and relying on agrochemicals that poison the Land.[2] As the industry struggles to maintain business as usual, what will that mean for the future of its workforce?

As the climate crisis worsens, the habitability of this region is dependent on the largely immigrant *trabajadores de la tierra* (land workers) who will do the critical restorative work—vegetation management, fire mitigation—to rehabilitate the Land. Many of the farmworkers who tend the vineyards of Sonoma County come from generations of land workers and carry with them a deep ancestral wisdom about how to live in right relationship with place. This ancestral knowledge—coupled with their lived experiences of tending the Land here in the North Bay—situates immigrant, Indigenous farmworkers as the real leaders in the fight to address the local impacts of climate change.

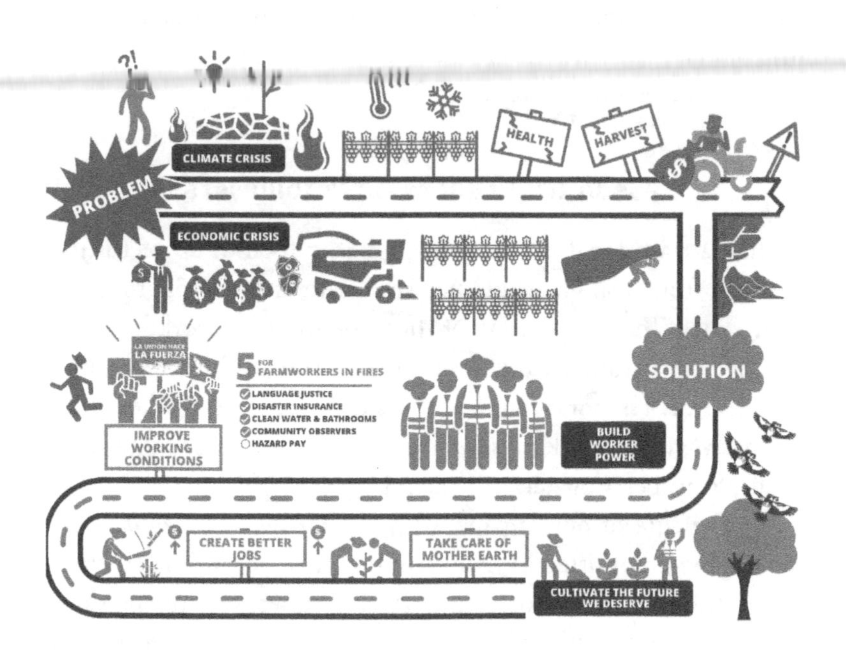

Fighting the Bad While Building the Good

At North Bay Jobs with Justice, we have found a potent organizing alchemy by fighting the bad while building the good: waging campaigns against the agrarian elite while empowering farmworkers to take leadership in climate resiliency work. We build power through structure-based organizing, and the strength and scale of our worker-led structures have allowed us to successfully fight to improve current jobs and create new, better jobs in vegetation management.

Over the past year and a half, four core lessons have emerged from our work: (1) worker leadership is critical to empowering a broad base of workers, (2) direct public action is necessary for workers to tap into their collective power, (3) the opposition is more easily outmaneuvered when worker leadership is flexible with its focus and willing to experiment with new tactics, and (4) patience and trust are critical for the challenging work of creating worker-led workforce models.

Lesson #1: Leaders have followers.

Unlike a traditional union campaign that focuses on recruiting workers from a single employer, we recruit from across multiple companies. The composition of the wine industry workforce creates the opportunity for this approach. The workforce is a mix of full-time, temporary visa, and resident seasonal workers who themselves could work for several employers or across vineyards of multiple property owners in a calendar year.

At the core of our organizing is the farmworker *equipo*—an organizing committee initially composed of seven leaders from across the industry who recruit other farmworkers to the campaign. The lesson we take from union organizing—where you need to get a majority of workers in a workplace to take collective action—is that *leaders have followers*. Rather than recruit any interested farmworker to join the equipo, we focus on recruiting leaders. A leader is someone who can get a group of people to act—to bring them to a meeting, picket line, or other direct action even when they might be scared. We prioritize organizer training so that workers can lead—not just in their workplaces but also in the broader community. Leaders may be responsible for organizing their direct coworkers as well as family members who work at other companies.

Leadership identification from broader kinship networks stemming from a shared region of origin and native language has been especially successful, allowing us to build a strong base among Oaxacan Indigenous communities like Chatino, Mixteco, Triqui, Maya, Mam, Zapoteco, and Tzotzil. For example, Maria Salinas is a focal point of the Chatino community in Sonoma County. In recent years, she has led efforts to preserve the Chatino language. In a classroom near the barn that serves as a community hall for many Chatino families, Maria teaches Chatino to children in the community. Maria's commitment to language justice motivated her to recruit Indigneous workers in her community to the 5 for Farmworkers campaign.

> Nuestros ancestros se fueron con el dolor de no ser escuchados. Yo sigo luchando porque no quiero que la siguiente generación siga sufriendo. (Our ancestors left with the pain of not having their voices heard. I keep fighting because I don't want the next generation to suffer what we've gone through.)[3]

Good organizing starts with meeting workers where they're at and listening and building authentic relationships. For the first half of 2021, we worked with worker-leaders like Maria to develop and conduct one hundred surveys with other workers about their experiences during wildfires. In June 2021, thirty farmworkers attended our *concilio* (worker council) to finalize the 5 for Farmworkers demands. As we moved to go public with the demands, our base expanded dramatically—led by the efforts of worker-leaders like Maria. Three concilios and four months later, over 150 farmworkers—plus 100 community supporters—picketed Simi Winery, owned by Constellation Brands, during a $150-a-plate harvest dinner. For many workers, it was their first picket line, and they showed up because worker-leaders like Maria knew how to get them there.

Lesson # 2: If the opposition only understands symphony, jazz becomes your strategy.

We framed the 5 for Farmworkers campaign as expansively as possible so that we didn't limit ourselves to any one sphere of engagement. The five demands made clear that everyone has a responsibility to improve farmworkers' conditions: wine producers, vineyard management companies, even local government. As a result, we could use a diversity of tactics and focus on different decision-makers as the opportunity arose. This allowed us to stay one step ahead of the opposition, which really began to mobilize only five months after the Simi Winery picket in the spring of 2022.

After the Simi Winery picket, we shifted our focus to the Sonoma County Board of Supervisors, which had the power to implement the 5 for Farmworkers demands through a formal evacuation zone access policy. Since the 2017 Tubbs Fire, the wine industry has relied on an ad hoc "ag pass" program that allowed employers to conduct harvests in evacuation zones without oversight or accountability. By focusing the fight on the adoption of a countywide policy, we created a clear choice for wine growers and elected officials to either support or oppose rules for farmworker safety during wildfires. As a result, we were able to get the first buy-in for our larger campaign from small organic growers who understand that the climate crisis requires new practices. We also gave sympathetic elected officials a clear path to take action.

Changing our target also confused our opposition, which was still waging an anti-union campaign even though we were in a policy fight. A handful of companies with leadership in the Sonoma County Winegrowers—a marketing commission that developed an astroturfing front group called Sonoma WISE—threw hundreds of thousands of dollars at union-busting experts and PR spokespeople. They failed to influence the policy in any significant way, but they did create prime organizing opportunities for our campaign. During the spring and summer of 2022, the Winegrowers brought managers and 150 paid H-2A (temporary visa) guest workers to a series of county board meetings in a theatrical attempt to display worker opposition to stronger protections. The strategy backfired. At those meetings, our worker-leaders talked to the workers about the campaign—resulting in dozens of new contacts with workers who reached out to us in subsequent weeks to report workplace abuses. The opposition's blunder also resulted in a Department of Labor investigation against two of the companies for violating H-2A contracts rules.

The script the Winegrowers deployed at county board meetings frequently pulled from an anti-union playbook that was completely unrelated to the policy at hand, with pro-employer managers claiming opposition because of an unwillingness to pay dues to an outside organization. (Though we certainly believe every worker deserves a union, North Bay Jobs with Justice isn't a union, and we don't require workers to pay dues.)

The start of harvest created the opportunity to also take action at the point of production in the vineyards. Months before, we had created a parallel structure to the farmworker equipo, training community members as *leaders* who could recruit and train crews of community safety observers (CSOs). As a result, nearly every night for over two months, we were able to send out a committed group of over thirty community members to gather data on working conditions during harvest. One of the demands of the 5 for Farmworkers was the institution of CSOs. Community allies took it upon themselves to make this demand a reality, even without a county policy or permission from the industry. The Winegrowers were furious. On more than one occasion, management called sheriff deputies in an attempt to intimidate and dissuade CSOs from continued action. In the final board of supervisors meeting to vote on the evacuation zone policy, the president of the Sonoma Winegrowers spent her entire public comment railing against CSOs instead of addressing the proposed policy. For vineyard managers who had gotten used to the impunity of operating at night—without outside observation—the consistent presence of community members gathering data was an unprecedented demonstration of community solidarity with workers.

In the policy fight, farmworkers won. In August 2022, despite reluctance by some supervisors, the board adopted for the first time a formal policy on ag access in evacuation zones with language on three of the 5 for Farmworkers demands: language justice, disaster insurance, and clean water and bathrooms. For frontline workers who lost out on income from evacuation zones, a first-of-its-kind $3 million disaster insurance fund for workers was established. And for Indigenous workers who had long experienced disrespect and discrimination, employers were now required to provide safety trainings in workers' preferred languages, including Indigenous languages.

Lesson #3: Direct action is the oxygen.

Throughout 2022, direct action provided a constant infusion of energy into the campaign while building up workers' confidence and assertiveness. We flyered at elegant tasting rooms, held pickets in front of county administration buildings, and hosted guerilla mural making, street theater demonstrations, and, to top off the year, a musical action in which workers blared banda music on the grounds of a chic musical festival at BR Cohn Winery, owned by Vintage Wine Estates.

Collective action invites workers to assert a collective public identity—deepening their involvement in the organizing drive. While flyering or picketing, workers and community supporters cannot be passive spectators: shared chants, songs, and movement all invite active participation and the forging of a collective public identity. Direct action is the bedrock of collective memory—a moment when a group of workers occupy a space in history as its protagonists. Anabel Garcia, a worker-leader with the campaign, explained it this way:

> A los ricos no les importa que tú has dejado tu vida en el fil haciendo las ganancias de ellos. Esta es la manera para hacer que nos vean y nos escuchen. (The rich don't care even after you've given your whole life to the fields making their profits. This is the way to make them see us and listen to us.)[4]

Many workers feel apprehensive the first time they take public action. When we took action against the BR Cohn Winery in October 2022, many of our new members were nervous. We had done many actions up to that point, but none as disruptive as driving up into a festival parking lot and hopping out with the horns blaring. Our members became especially nervous when sheriff deputies rolled up. But moving through that nervousness together—and experiencing the joy of collective action on the other side—brought everyone closer together, strengthening the campaign in the long run.

For an industry like wine, which is built on brand image and marketing, even the specter of direct public action can be leveraged for immediate workplace victories. In 2022, workers won back over $50,000 in unpaid wages from their employers by using the specter of direct action as leverage. For this to work, the likelihood that we will take action needs to be credible. After a year of continued direct action under the NBJwJ banner, employers have a good sense of what they want to avoid. As a nontraditional labor organization, NBJwJ doesn't have the institutional power of collective bargaining agreements or open-ended legal action, so we have to rely on direct action to get the goods. But doing so means that we empower workers as agents in their own victories—helping to build our organizational strength and power. It also gives workers the opportunity to assert their agency in making a victory happen.

Lesson #4: Utopia doesn't build itself, and building it isn't easy.

In collaboration with organizations like NBJwJ, Resilience Force, Russian Riverkeeper, and Occidental Arts & Ecology Center, farmworker-leaders have begun to build models of land stewardship that center worker leadership and traditional ecological knowledge. In 2022, our organizing team recruited and coordinated crews of farmworkers from the 5 for Farmworkers campaign to lead five vegetation management pilot projects. The projects leveraged public funds to pay workers nearly double what they would have made in the vineyards. With millions of contract dollars pouring into the county to do this climate resiliency work, organized workers are building a just transition from a waning wine industry on their own terms.

During the first month-long pilot project in January 2022, the sounds of the cuadrilla echoed under the redwood and fir canopy as workers masterfully wielded machetes to clear dead tan oak and overgrown vegetation. During a second project in August 2022, in collaboration with Russian Riverkeeper, the crew removed invasive Arundo, or as the workers call it *carrizo*, a fifteen-foot-tall reed similar to bamboo that grows along waterways, sucking up a disproportionate amount of water and crowding out native plants. Many had been in relationship with this plant in their homelands: it was the material used for first cribs, baskets that bring chiles to market, and decorations for weddings. Workers took great pride in their knowledge—leaving the worksite with a few poles of carrizo to build structures for beans and tomatoes in their home gardens or finishing the project with mezcal in a cup made from the plants they had cleared.

The cuadrilla creates models of work without fixed hierarchies or bosses present, but the "bosses in our head" often create new problems. For example, for farmworkers who have spent significant parts of their lives in that industry, the switch to a slower, more intentional pace of work can be challenging. At times, the transition created tension among workers who felt that other crew members weren't pulling their weight.

Tension is inevitable and, to some degree, necessary to a healthy structure. But engaging in hard conversations requires patience and trust. The core work crew, or cuadrilla, on the projects was made up of worker-leaders who came out of the 5 for Farmworkers campaign, meaning workers started from a place of trust and cohesion around a shared politicized struggle. Early on, to help navigate tension, the cuadrilla decided that it would make the most sense to elect a *responsable* who would, for that project, hold the primary leadership responsibilities, making sure that crew members were taking the necessary breaks and working at a pace that honored the needs of everyone on the team. However, the responsable was not getting paid more than the rest of the crew and still had to perform the same work duties. To create a democratic space, the role rotated among crew members at the start of a new project.

The workers were building their own structures based on mutual respect and hard work. For Sandra De Leon, a worker-leader with the campaign, this represented a radical redefinition of work.

> En los viñedos es mucha presión para sacar el trabajo. Entre más hagamos nosotros, más ganen los patrones. Pero aquí, trabajamos en nuestro paso y tenemos el derecho a opinar. Como compañeros estamos cuidando uno a otro y nos estamos poniendo en acuerdo cómo hacer el trabajo. (In the vineyards, there is a lot of pressure to get the job done. The more we do, the more the bosses earn. But here, we work at our pace, and we have the right to have an opinion. As compañeros, we are taking care of each other and deciding together how to do the job.)[5]

Having worker-leaders who are involved in both the 5 for Farmworkers campaign and the climate resilience work creates opportunities for political education that is actualized in practice. NBJwJ organizers frequently spend time out in the field doing the resilience work alongside the cuadrilla—like hacking down carrizo—learning from worker-leaders and deepening trust. We move between talking through experiences with abusive bosses to sharing ecological knowledge between machete swings. The politicized collective struggle is lived out and returned to—both on the picket line and in caring for the Land.

What Comes After the Fires?

Rather than being satisfied with just a bigger piece of a shrinking pie, farmworkers are setting out to create something entirely different. Amidst growing climate chaos, our future will be determined by the leadership of immigrant and Indigenous workers who will do the critical work the Land desperately needs. This means not just fighting to improve current jobs but beginning to actualize a vision of worker-led stewardship rooted in real relationship with the Land.

When farmworkers reclaim the creative power of their own labor and tap into their own reservoirs of knowledge and experience, they are tapping into something powerful. As summed up by Margarita Garcia, a proud Mixteca campesina and worker-leader,

> Estamos luchando no solamente para recoger las uvas para hacer más rico a los ricos, sino estamos aquí porque amamos a la tierra y amamos nuestra cosecha. (We're not fighting to just keep picking grapes to make the rich richer—no, we're here fighting because we love the Land and we love our harvest.)[6]

What comes after the fires? On the charred slopes of the Mayacamas Mountains, between the burnt remains of what was, you'll find new growth. Manzanita regrowing from sturdy roots, young knobcone pines, whose seeds are released only by flames, wildflowers in full bloom. Regeneration. Amidst crisis, the Land reminds us that while change is inevitable, the seeds of resurgent possibilities are ready to emerge from the ashes.

ACKNOWLEDGMENTS

So many brilliant organizers took time out of their packed days to write these articles. This book would not have been possible without their dedication to fighting for economic and climate justice—and to sharing their strategies so that we can build the biggest movement possible. Miya Yoshitani and Shantell Bingham played an especially significant role in shaping this anthology, offering much-needed advice, inspiration, and connections with contributors. Matt Mayers was equally helpful within the House of Labor, providing connections and engaging in significant follow up. Dania Rajendra provided insight and connections on the coalitional work to fight Amazon. Jeff's participation in an It Takes Roots and Climate Justice Alliance delegation to Glasgow in 2022 was also critical in introducing him to some of the key thinkers in this space.

Katey Lauer first introduced Lindsay to Julie Enszer—then at The New Press—to talk about books on organizing. Julie was a fantastic sounding board and champion of this anthology when it was still just an idea. Ben Woodward took over early in the project, and has been an insightful editor and invaluable collaborator at every step of the process. Ishan Desai-Geller, Emily Albarillo, and the entire team at The New Press have shepherded the anthology through to completion with attentiveness and care.

This book started as a project of *The Forge: Organizing Strategy and Practice*, where Lindsay was working as the editor. Many thanks to Brian Kettenring and the entire *Forge* Editorial Advisory Board and Publishing Committee for creating such a vibrant space for organizers to share strategies with each other—and for supporting Lindsay in developing this anthology as one piece of that work.

Both of us are honored to be surrounded by a generous and caring movement community, without whom we would not have been able to put this anthology together. You are too numerous to mention, but we love you and feel you at our backs.

Most importantly, this anthology would not exist if it were not for the courage of workers who grow our food, build our infrastructure, and power our society—often at great risk to themselves. Their willingness to stand up to the boss to improve their own working conditions and to fight for a more just future for us all has made so much possible. We see you, and we hope that this anthology can play a small part in creating the revolutionary labor and environmental justice movement you need and deserve.

NOTES

Preface

1. Slogan appearing on Walter Crane's *The Workers' Maypole* cartoon, published in 1894. See Laura Foster, "Radical Object: Walter Crane's The Workers' Maypole (1894)," *History Workshop*, May 1, 2020.

2. Ryan Grim, "The Railroad Strike Was the Product of Eight Years of Militant Rank-and-File Organizing," *The Intercept*, December 11, 2022.

3. Jane McAlevey, "The West Virginia Teachers Strike Shows That Winning Big Requires Creating a Crisis," *The Nation*, March 12, 2018.

4. Alyssa Battistoni, "Living, Not Just Surviving," *Jacobin*, August 15, 2017.

5. Mindy Isser, "The Green New Deal Just Won a Major Union Endorsement. What's Stopping the AFL-CIO?" *In These Times*, August 12, 2020.

6. See, for example, Labour Party, "It's Time for Real Change: The Labour Party Manifesto 2019," labour.org.uk/wp-content/uploads/2019/11/Real-Change-Labour-Manifesto-2019.pdf.

7. Leigh Phillips, "Blue Collars, Green Jobs?" *Breakthrough Institute*, November 30, 2021.

8. Tony Mazzocchi, "A Superfund for Workers?" *Earth Island Journal* 9, no. 1 (Winter 1993/94): 40–41.

9. Steven P. Nesbit and Lonnie R. Stephenson, "Nuclear Energy Ensures Clean Energy Jobs for American Workers," *Nuclear Newswire*, September 8, 2021.

10. Holly Jean Buck, *Ending Fossil Fuels: Why Net Zero Is Not Enough* (London: Verso, 2021).

11. Naomi Klein, *This Changes Everything: Capitalism vs. the Climate* (New York: Simon and Schuster, 2014).

Introduction

1. Colin Sullivan and Debra Kahn, "Voters Reject 2-Sided Assault on Climate Law," *New York Times*, November 3, 2010.

2. Ina Jaffe, "Oil Firms Bankroll Calif. Climate-Change Measure," NPR, November 1, 2010.

3. Melissa Lentz, "Activists Work to Shut Down Tar Sands Oil Pipelines," Fox 21, October 11, 2016, www.fox21online.com/2016/10/11/activists-work-to-shut-down-tar-sands-oil-pipelines.

4. Tony Mazzocchi, "A Superfund for Workers?," *Earth Island Journal* 9, no. 1 (Winter 1993/94): 40–41.

5. Manuel Pastor, *Still Toxic After All These Years: Air Quality and Environmental Justice in the San Francisco Bay Area* (Santa Cruz: Center for Justice, Tolerance, and Community, University of California, Santa Cruz, 2007); Jane Kay and Cheryl Katz, "Pollution, Poverty, and People of Color: The Factory on the Hill," *Environmental Health News*, June 4, 2012, www.ehn.org/pollution-poverty-richmond-2645503359.html; Steve Early, *Refinery Town: Big Oil, Big Money, and the Remaking of an American City* (Boston: Beacon Press, 2017).

1. The Dream and the Nightmare

1. Los Angeles County, "CSO Current Initiatives," www.cso.lacounty.gov/the-plan/cso-current-initiatives/.

2. Kirsten Korosec, "Ford to Invest $3.7B in US Factories, Add 6,200 Union Jobs in Push to Build More EVs," *Tech Crunch*, June 2, 2022, techcrunch.com/2022/06/02/ford-to-invest-3-7b-in-u-s-factories-add-6200-union-jobs-in-push-to-build-more-evs/; Neal E. Boudette, "Ford Will Build 4 Factories in a Big Electric Vehicle Push," *New York Times*, September 27, 2021; Rachel Parkes, "Delta Places Huge Order for Sustainable Aviation Fuel Made with 839mw of Green Hydrogen," *Recharge News*, September 6, 2022, rechargenews.com/energy-transition/delta-places-huge-order-for-sustainable-aviation-fuel-made-with-839mw-of-green-hydrogen/2-1-1291711; Michael Tyrrell, "Air France-KLM and Moresand Sign SAF Agreement," *Aero-mag*, January 11, 2023, aero-mag.com/air-france-klm-and-moresand-sign-saf-agreement.

3. Dave Kunz, "Volvo Introduces New Electric Cars but Gas Models Still Available for Now," ABC7, January 5, 2023, abc7.com/volvo-electric-vehicles-hybrid-cars-gas-engine/12653932; Matt O'Leary, "Volvo Group and

Fossil-Free Steel," *Volvo Group*, June 1, 2022, www.volvogroup.com/en/news-and-media/news/2022/jun/volvo-group-and-fossil-free-steel.html.

4. Neal E. Boudette and Coral Davenport, "G.M. Will Sell Only Zero-Emission Vehicles by 2035," *New York Times*, January 28, 2021.

5. California Air Resources Board, "2022 Scoping Plan for Achieving Carbon Neutrality," November 16, 2022, ww2.arb.ca.gov/sites/default/files/2022-12/2022-sp.pdf.

6. California Energy Commission, "California's Oil Refineries," California State Portal, February 7, 2023, www.energy.ca.gov/data-reports/energy-almanac/californias-petroleum-market/californias-oil-refineries.

3. Pushing for a Green New Deal for Education from Below

1. Joseph B. Treaster and Abby Goodnough, "Powerful Storm Threatens Havoc Along Gulf Coast," *New York Times*, August 29, 2005.

2. Emery Winter, "Yes, Rare Extreme Weather Events Are Happening More Frequently," *Verify This*, September 8, 2021, www.verifythis.com/article/news/verify/extreme-weather-verify/extreme-weather-events-100-year-floods-storms-wildfires-more-frequent-often/536-3352b5ca-3b72-4215-8421-194dd761f40a.

3. Meghan Gallagher, "When Climate Change Forces Schools to Close: Fires, Storms and Heatwaves Have Already Kept 1 Million Students out of Classrooms This Semester," *The 74 Million*, September 16, 2021, www.the74million.org/article/when-climate-change-forces-schools-to-close-fires-storms-and-heatwaves-have-already-kept-1-million-students-out-of-classrooms-this-semester.

4. Rebecca K. Miller and Iris Hui, "Impact of Short School Closures (1–5 Days) on Overall Academic Performance of Schools in California," *Scientific Reports* 12, no. 2079 (2022), www.nature.com/articles/s41598-022-06050-9.

5. For more detail on educational impacts of extreme weather, see Rajashi Chakrabarti, "The Impact of Superstorm Sandy on New York City School Closures and Attendance," *Huffington Post*, December 24, 2012.

6. Substance Abuse and Mental Health Services Administration, "Disaster Technical Assistance Center Supplemental Research Bulletin Behavioral Health Conditions in Children and Youth Exposed to Natural Disasters," September 2018, www.samhsa.gov/sites/default/files/srb-childrenyouth-8-22-18.pdf.

7. American Federation of Teachers, "In Support of Green New Deal," Union resolution, 2020, www.aft.org/resolution/support-green-new-deal.

8. American Federation of Teachers, "Green New Deal."

9. American Federation of Teachers, "Divest from Fossil Fuels and Reinvest in Workers and Communities," Union resolution, 2022, www.aft.org/resolution/divest-fossil-fuels-and-reinvest-workers-and-communities.

10. For more on GNDs from below, see Jeremy Brecher, "Unions Making a Green New Deal from Below—Part 1," *Labor Network for Sustainability*, 2022, www.labor4sustainability.org/strike/unions-making-a-green-new-deal-from-below-part-1/.

11. Andrew Bauld, "Why Schools Need to Look at Their Own Carbon Footprint," Harvard Graduate School of Education, *Usable Knowledge*, November 1, 2021, www.gse.harvard.edu/news/uk/21/11/why-schools-need-look-their-own-carbon-footprint.

12. An initial project could be the installation of building-level metering to track energy use, emissions, water use, and more, in order to track progress toward meeting sustainability goals.

13. For additional green school ideas, see Akira Drake Rodriguez, Daniel Aldana Cohen, Erika Kitzmiller, Kira McDonald, David I. Backer, Neilay Shah, Ian Gavigan et al., "A Green New Deal for K-12 Public Schools," *Climate + Community Project*, July 2021, www.climateandcommunity.org/gnd-for-k-12-public-schools; K12 Climate Action Commission, "K-12 Climate Action Plan," *Aspen Institute*, September 2021, www.thisisplaneted.org/img/K12-ClimateActionPlan-Complete-Screen.pdf.

14. U.S. Department of the Treasury, "Fact Sheet: Four Ways the Inflation Reduction Act's Tax Incentives Will Support Building an Equitable Clean Energy Economy," 2022, home.treasury.gov/system/files/136/Fact-Sheet-IRA-Equitable-Clean-Energy-Economy.pdf.

15. Winning support from local building trades unions to support Green Schools projects is an essential part of a successful organizing campaign to win support for these investments.

16. See, e.g., "CJA and the Green New Deal: Centering Frontline Communities in the Just Transition," Climate Justice Alliance, climatejusticealliance.org/gnd/.

17. "'Our House Is on Fire,' Climate Justice Resolution by Directors Peterson

and Taylor," *Rethinking Schools*, February 27, 2020, rethinkingschools.org/wp-content/uploads/2020/10/Climate_resolution_peterson_2_27_20.pdf.

18. Learn more, including how to get involved, by contacting LNS through their website, labor4sustainability.ourpowerbase.net/ECAN.

4. Care Work Is Central to a Just Transition

1. "2017 Incident Archive," Cal Fire Department of Forestry and Fire Protection, www.fire.ca.gov/incidents/2017/.

2. "Kaiser, Sutter Santa Rosa RNs Help Evacuate Wildfire Patients, While Also Facing Personal Losses," *National Nurses United*, December 6, 2017, www.nationalnursesunited.org/press/kaiser-sutter-santa-rosa-rns-help-evacuate-wildfire-patients-while-also-facing-personal.

3. Yonah Freemark, Billy Fleming, Caitlin McCoy, Rennie Meyers, Thea Riofrancos, Xan Lillehei, and Daniel Aldana Cohen, "Toward a Green New Deal for Transportation: Establishing New Federal Investment Priorities to Build Just and Sustainable Communities." *Climate + Community Project*, 2022; Akira Drake Rodriguez, Daniel Aldana Cohen, Erika Kitzmiller, Kira McDonald, David I. Backer, Neilay Shah, Ian Gavigan, Xan Lillehei, A. L. McCullough, Al-Jalil Gault, Emma Glasser, Nick Graetz, Rachel Mulbry, and Billy Fleming. "Transforming Public Education: A Green New Deal for K–12 Public Schools," *Climate + Community Project*, 2021, www.climateandcommunity.org/gnd-for-k-12-public-schools.

4. Alyssa Battistoni, "Ways of Making a Living: Revaluing the Work of Social and Ecological Reproduction," *Socialist Register* 56 (2020).

5. Bureau of Labor Statistics, U.S. Department of Labor, "Occupational Outlook Handbook, Home Health and Personal Care Aides," www.bls.gov/ooh/healthcare/home-health-aides-and-personal-care-aides.htm.

6. Mina Kim and Lakshmi Sarah, "'The Work That Makes All Other Work Possible': Ai-Jen Poo on Why Home Care Workers Are Infrastructure Workers," KQED, n.d., accessed December 7, 2022, www.kqed.org/news/11877838/the-work-that-makes-all-other-work-possible-ai-jen-poo-on-why-home-care-workers-are-infrastructure-workers.

7. Barbara Ehrenreich and Arlie Russell Hochschild, *Global Woman: Nannies, Maids, and Sex Workers in the New Economy* (New York: Henry Holt & Company, 2003).

8. Across the United States, in 2020, median wages for home care workers ranged from $9.05 to $16.66 per hour. U.S. Bureau of Labor Statistics, "Occupational Employment and Wages, May 2020: 31-1120 Home Health and Personal Care Aides," March 31, 2021, www.bls.gov/oes/current/oes311120.html.

9. Caitlin McLean, "Early Childhood Workforce Index 2020," Center for the Study of Child Care Employment, University of California, Berkeley, 2021, www.cscce.berkeley.edu/workforce-index-2020/wp-content/uploads/sites/2/2021/02/Early-Childhood-Workforce-Index-2020.pdf.

10. "Child Care Sector Jobs," Center for the Study of Child Care Employment, December 5, 2022, cscce.berkeley.edu/publications/brief/child-care-sector-jobs-bls-analysis/.

11. Rasheed Malik, Katie Hamm, and Leila Schochet, "America's Child Care Deserts in 2018," Center for American Progress, December 6, 2018, www.americanprogress.org/article/americas-child-care-deserts-2018/.

12. "Demanding Change: Repairing Our Child Care System," Child-Care Aware of America, February 2022, info.childcareaware.org/hubfs/FINAL-Demanding%20Change%20Report-020322.pdf.

13. Kim Parker and Jessica Horowitz, "Majority of Workers Who Quit a Job in 2021 Cite Low Pay, No Opportunities for Advancement, Feeling Disrespected," Pew Research Center, March 9, 2022, www.pewresearch.org/fact-tank/2022/03/09/majority-of-workers-who-quit-a-job-in-2021-cite-low-pay-no-opportunities-for-advancement-feeling-disrespected/.

14. United Teachers Los Angeles, "FAQ: The Fight for Black Lives," utla.net/resources/faq-the-fight-for-black-lives/.

15. "#StudentsDeserve," Students Deserve Coalition, www.schoolslastudentsdeserve.com/.

16. "The Beyond Recovery Platform—UTLA," United Teachers Los Angeles, utla.net/app/uploads/2022/07/Beyond-Recovery-Platform_Full.pdf.

17. "The Beyond Recovery Platform."

18. US Environmental Protection Agency, "Sources of Greenhouse Gas Emissions," www.epa.gov/ghgemissions/sources-greenhouse-gas-emissions.

19. Dana Goldstein and Elisabeth Dias, "Oklahoma Teachers End Walkout After Winning Raises and Additional Funding," *New York Times*, April 12, 2018.

20. Colin Carlson, Gregory Albery, Cory Merow, et al., "Climate

Change Increases Cross-Species Viral Transmission Risk," *Nature* 607 (2022): 555–562.

21. Adam Dean, Jamie McCallum, Simeon D. Kimmel, and Atheendar S. Venkataramani, "Resident Mortality and Worker Infection Rates from Covid-19 Lower in Union Than Nonunion US Nursing Homes, 2020–21," *Health Affairs* 41, no. 5 (2022): 751–59, doi.org/10.1377/hlthaff.2021.01687.

22. "The 2018 Report of the Lancet Countdown on Health and Climate Change: Shaping the Health of Nations for Centuries to Come," *The Lancet* 392 , no. 10163 (2018).

23. Dean et al., "Resident Mortality and Worker Infection Rates from Covid-19."

24. Iris Altamirano, Greg Nammacher, and Priya Dalal-Whelan, "Lessons from the First Union Climate Strike in the U.S.," *Labor Notes*, April 30, 2020, labornotes.org/2020/04/lessons-first-union-climate-strike-us.

25. Altamirano et al., "Lessons from the First Union Climate Strike in the U.S."

5. Young Workers Can Bridge the Labor and Climate Movements

1. Kathleen (not her real name) was interviewed as part of the Young Worker Listening Project on June 15, 2021.

6. Organizing Climate Jobs Rhode Island

1. Rhode Island Labor History Society, "2022 Saylesville Commemoration Speech by Erik Loomis," September 5, 2022, www.youtube.com/watch?v=gzv7EhOrRes&t=2042s.

2. Erik Loomis, "Why Labor and Environmental Movements Split—and How They Can Come Back Together," *Environmental Health News*, September 18, 2018, www.ehn.org/labor-and-environmental-movements-merge-2605763191.html.

3. As the historian Erik Loomis has written, "neither the labor movement nor the environmental movement has the power that it did four decades ago." Loomis, "Why Labor and Environmental Movements Split."

4. R.K. Sokas, X.S. Dong, and C.T. Cain, "Building a Sustainable Construction Workforce," *International Journal for Environmental Research and*

Public Health 16, no. 21 (2019): 4202, doi: 10.3390/ijerph16214202. PMID: 31671567; PMCID: PMC6862229.

5. You can find the report at https://climatejobsri.org/resources/.

10. Good Jobs, Clean Air

1. Tom Bergeron, "New Distinction for Amazon: Largest Employer in N.J.," ROJ-NY, September 15, 2020, www.roi-nj.com/2020/09/15/industry/new-distinction-for-amazon-largest-employer-in-n-j/.

2. Linda Linder, "Amazon Investment in New Jersey Area Surpasses $23B Over Past Decade," NJBIZ, August 11, 2021, njbiz.com/amazon-investment-in-new-jersey-area-surpasses-23b-over-past-decade/.

3. Karen Yi, "Newark Literally Sent Amazon a Huge Valentine's Day Heart to Try to Get Its Attention," NJ.com, February 15, 2019, www.nj.com/essex/2019/02/newark-literally-sent-amazon-a-huge-valentines-day-heart-to-try-to-get-its-attention.html; Michael Catalini, "New Jersey Promising Billions to Get Amazon Headquarters," NorthJersey.com, January 11, 2018, www.northjersey.com/story/news/new-jersey/2018/01/11/new-jersey-promising-billions-get-amazon-headquarters/1026592001/.

4. Natalie Kostelni, "Report Details Amazon's Multi-Billion Dollar Investment in Philadelphia Region, but What About the Bigger Picture?" *Philadelphia Business Journal*, August 11, 2021, www.bizjournals.com/philadelphia/news/2021/08/11/amazon-invested-billions-in-philadelphia-region.html.

5. Daniel Munoz, "New Report Details Injuries at NJ Amazon Warehouses as Activists Oppose New Facility," NorthJersey.com, April 28, 2022, www.northjersey.com/story/news/2022/04/28/nj-amazon-worker-injuries-twice-high-other-warehouses-report/9575360002.

6. Carmen Martino and Nicole Rodriguez, "As Injury Rates at Amazon Rise Nationwide, Amazon Workers in New Jersey Pay a Heavy Toll," *Research Brief*, 2021.

7. For examples of opposition to warehouse growth in New Jersey, see Marin Resnick, "Clinton Township Nixes Warehouse on Exxonmobil Property," *New Jersey Hills*, November 2, 2022, www.newjerseyhills.com/hunterdon_review/news/video-clinton-township-nixes-warehouse-on-exxonmobil-property/article_e298187a-0136-5895-ba95-e137c0432d2f.html; Jackie Roman, "N.J. to Issue Its First Guideline on Where Warehouses Should be Built," NJ.com, August 8, 2022, www.nj.com/news/2022/08/nj-to-issue-its-first-guidelines-on

-where-warehouses-should-be-built.html; Amy Goldsmith, "Warehouses Are Taking Over New Jersey. Stop Building Them Now," NJ.com, April 23, 2021, www.nj.com/opinion/2021/04/warehouses-are-taking-over-new-jersey-stop-building-them-now-opinion.html; Jerry Carino, "Garden State or Warehouse State? Jackson Project Could Spur Changes in NJ Law," NJ.com, April 26, 2021, www.app.com/story/news/local/communitychange/2021/04/26/nj-warehouse-state-jackson-project-could-spur-changes-state-law/7336436002/.

8. Tennyson Donyea, "Protesters Bash Trump's Raids and Amazon Saying 'No Tech for ICE,'" July 15, 2019, www.nj.com/news/2019/07/protestors-bash-trumps-raids-and-amazon-saying-no-tech-4-ice.html; Kenneth Burns, "Bellmawr Amazon Workers Walk Off the Job in Protest of Facility Transfers," WHYY, June 1, 2022, whyy.org/articles/bellmawr-amazon-walkout-warehouse-transfers/; Anna Schier, "Amazon Workers Walk Off the Drive in Avenel to Protest Wage Cuts," Patch, June 18, 2022, patch.com/new-jersey/woodbridge/amazon-drivers-walk-job-avenel-protest-wage-cuts; Make the Road New Jersey, "We're LIVE: It's Prime Day and Workers and Community Members join @maketheroadnj and Clean Water Action New Jersey," Facebook, October 13, 2020, www.facebook.com/maketheroadnj/videos/1282995545401825.

9. Carino, "Garden State or Warehouse State?"

10. Chris Fry, "Amazon to Invest $125 Million in Air Cargo Facility at Newark Airport," *Jersey Digs*, September 7, 2021, jerseydigs.com/amazon-to-invest-125-million-in-air-cargo-facility-at-newark-airport/.

11. "Port Authority Board Authorizes Agency to Enter into 20 Year Lease with Amazon Global Air to Redevelop and Operate World Class Air Cargo Campus at Newark Liberty Airport," press release, August 5, 2021, www.panynj.gov/port-authority/en/press-room/press-release-archives/2021-press-releases/port-authority-board-authorizes-agency-to-enter-into-20-year-lease-with-amazon-glob,l-air.html; "Hedge Papers No. 74: Dirty Deal: How the Port Authority's Backroom Deal with Amazon Would Harm Black and Brown Communities in Newark," October 2021, www.populardemocracy.org/sites/default/files/Hedge Clippers Amazon 'Dirty Deal' Port Authority Report October 2021.pdf; Max Rivlin-Nadler, "Hell on Wheels: Port Authority's Broken Promise Is Choking Newark Kids," *Village Voice*, May 3, 2016, www.villagevoice.com/2016/05/03/hell-on-wheels-port-authoritys-broken-promise-is-choking-newarks-kids/.

12. "Hedge Papers No. 74."

13. "Advocates Protest Amazon Air Cargo Hub Coming to Newark Airport," NorthJersey.com, October 6, 2021, www.northjersey.com/videos/news/essex/2021/10/06/advocates-protest-amazon-air-cargo-hub-coming-newark-airport/6024881001/.

14. Noam Scheiber and Karen Weise, "Amazon Hub in Newark Is Canceled After Unions and Local Groups Object," *New York Times*, July 7, 2022.

15. Mary Ann Koruth, "Worker Advocates Demand Transparency in Port Authority's Million-Dollar Deal with Amazon," North Jersey, October 6, 2021, www.northjersey.com/story/news/2021/10/06/amazon-newark-nj-worker-advocates-airport-hub/6003511001/; "Hedge Papers No. 74."

16. Carmen Martino and Nicole Rodriguez, "As Injury Rates at Amazon Rise Nationwide, Amazon Workers in New Jersey Pay a Heavy Toll," *ResearchBrief*, 2021.

17. Good Jobs Clean Air NJ, "Amazon's Growth in New Jersey and Falling Wages in the Delivery and Warehouse Industry," March 2022, drive.google.com/file/d/1vlBqUQtU9KR12PzT5c2eZ9AIVr3UG2Ek/view.

18. "Annual Worker Turnover at Amazon Warehouses in New Jersey Is Approximately 124 Percent, Almost Double the Rate of Turnover at Non-Amazon Warehouses in the State," National Employment Law Project, June 2022, www.nelp.org/publication/dead-end-jobs-amazon-warehouses-fail-to-provide-long-term-full-time-employment-for-new-jerseyans/.

19. Daniel Munoz, "New Report Details Injuries at NJ Amazon Warehouses as Activists Oppose New Facility," NorthJersey.com, April 28, 2022, www.northjersey.com/story/news/2022/04/28/nj-amazon-worker-injuries-twice-high-other-warehouses-report/9575360002/; Sophie Nieto-Munoz, "Excessive Injuries, Low Pay at Amazon Warehouses Cause High Turnover, Report Claims," *New Jersey Monitor*, June 23, 2022, newjerseymonitor.com/2022/06/23/excessive-injuries-low-pay-at-amazon-warehouses-cause-high-turnover-report-claims; Larry Higgs, "Protesters Oppose Newark Airport Amazon Air Hub Deal at Port Authority's First In-Person Meeting," NJ.com, May 17, 2022, www.nj.com/news/2022/03/protesters-oppose-newark-airport-amazon-air-hub-deal-at-port-authoritys-first-in-person-meeting.html.

20. Star-Ledger Editorial Board, "Are Injuries Surging at Amazon? The Feds Should Investigate | Editorial," NJ.com, June 16, 2022, www.nj.com/opinion/2022/06/are-injuries-surging-at-amazon-the-feds-should-investigate-editorial.html.

21. "Rep. Norcross Calls for OSHA Investigation into Amazon Warehouses After Reports of Skyrocketing Injuries," Donald Norcross: Working for New Jersey, May 9, 2022, www.donaldnorcrossforcongress.com/media/norcross-osha-investigation-amazon-warehouses/; "OSHA Opens Second Amazon Probe Following Two More Worker Deaths in New Jersey," CBS News, August 12, 2022, www.cbsnews.com/news/amazon-worker-deaths-rafael-reynaldo-mota-frias-osha-investigation-new-jersey/.

22. Sophie Nieto-Munoz, "Local Officials Argue Against Amazon Deal at Newark Airport," *New Jersey Monitor*, February 16, 2022, newjerseymonitor.com/2022/02/16/local-officials-argue-against-amazon-deal-at-newark-airport.

23. @MaketheRoadNewJersey, Twitter, April 28, 2022, twitter.com/MaketheRoadNJ/status/1519811569747644417.

24. Larry Higgs, "Port Authority Asked to End 'Secret' Negotiations for Newark Airport Amazon Air Freight Hub," NJ.com, June 21, 2022, www.njb.com/news/2022/06/port-authority-asked-to-end-secret-negotiations-for-newark-airport-amazon-air-freight-hub.html.

25. Donald Payne Jr., "Congressman to Amazon: Stop Battling the Unions If You Want to Build at Newark Airport," NJ.com, May 11, 2022, www.nj.com/opinion/2022/05/congressman-to-amazon-stop-battling-the-unions-if-you-want-to-build-at-newark-airport-opinion.html.

26. Noam Scheiber, "A Union Blitzed Starbucks. At Amazon, It's a Slog," *New York Times*, May 12, 2022.

27. Bill Pascrell Jr., Donald Norcross, Tom Malinowski, Donald M. Payne Jr., Frank Pallone, Mikie Sherril, and Bonnie Watson Coleman, Letter to Douglas Parker, May 25, 2022, norcross.house.gov/_cache/files/e/2/e211b709-dee5-4241-b1d6-a4515aa1862c/42FF139D2F36B0298D06CEE1146A1D36.220525—-norcross-osha-amazon-letter.pdf.

28. Sophie Nieto-Munoz and Dana Difilippo, "Congressmen Call for Amazon Probe After Worker's Death," *New Jersey Monitor*, July 28, 2022, newjerseymonitor.com/2022/07/28/congressmen-call-for-amazon-probe-after-workers-death/.

29. Kate Briquelet and Josh Fiallo, "Amazon Employee Who Died on Prime Day Was Hardworking Dad," *Daily Beast*, July 26, 2022, www.thedailybeast.com/amazon-employee-who-died-on-prime-day-rafael-reynaldo-mota-frias-was-hardworking-dad.

30. Briquelet and Fiallo, "Amazon Employee Who Died."

11. Killing the Wiindigo

1. Just Transition AK, "Kohtr'elneyh—Remembering Forward: A Strategic Framework for a Just Transition," drive.google.com/file/d/16Yh9645GuHj_sQhu-9PGMxY9d1MAaM4k/view..

2. Austin Stagman, "How Strong Is the U.S. Consumer?," *Inside Indiana Business*, May 8, 2023, www.insideindianabusiness.com/articles/how-strong-is-the-u-s-consumer.

3. "Growing Consumption," Knowledge for Policy, February 16, 2023, knowledge4policy.ec.europa.eu/growing-consumerism_en; United Nations (UN), "Global Population," *United Nations*, www.un.org/en/global-issues/population.

4. Dr. Patrick Mooney, executive director of ETC Canada, delivered these remarks in Session Five, Supporting Agrobiodiversity, November 4, 2015, Indigenous Terra Madre, Shillong, Meghalaya, India.

5. Patrick Mooney and Rural Advancement Foundation International, *The Parts of Life: Agricultural Biodiversity, Indigenous Knowledge, and the Role of the Third System* (Uppsala: Dag Hammarskjöld, 1996).

6. "Climate-Smart Agriculture," World Bank, April 5, 2021, www.worldbank.org/en/topic/climate-smart-agriculture.

7. Rebekah Clarke, "Fast Fashion's Carbon Footprint," Carbon Literacy, August 2021, carbonliteracy.com/fast-fashions-carbon-footprint.

8. Valerie Vande Panne, "Fashion Is Quietly a Major Fossil Fuel Industry—Could a Sustainable Revolution Be Around the Corner?," *Salon*, September 6, 2019, www.salon.com/2019/09/06/fashion-is-quietly-a-major-fossil-fuel-industry-could-a-sustainable-revolution-be-around-the-corner/.

9. "Solar Thermal Heating," Lightspring Solar, www.lightspring.io/solar thermal.html.

12. "We Are Our Best Chance for Rescue"

1. Lauren Kaori Gurley, "Shifting America to Solar Power Is a Grueling, Low-Paid Job," *Vice*, June 27, 2022, www.vice.com/en/article/z34eyx/shifting-america-to-solar-power-is-a-grueling-low-paid-job.

2. U.S. Bureau of Labor Statistics, "Top 10 Fastest Growing Occupations, Excluding Pandemic Recovery," September 8, 2021.

3. American Clean Power, "2021 Clean Energy Labor Supply," 8.

4. Robert Pollin, Chirag Lala, and Shouvik Chakraborty, "Job Creation Estimates Through Proposed Inflation Reduction Act," Political Economy Research Institute at University of Massachusetts Amherst, August 2022.

5. GWA staff email communication with Alicia Ramirez, October 16, 2022.

6. Gurley, "Shifting America to Solar Power."

7. Trent Nylander, "The Dangers of Being a Wind Technician," Green Workers Alliance, July 21, 2022, https://www.greenworkers.org/post/the-dangers-of-being-a-wind-technician.

8. Brittney Linton remarks, GWA reception, Washington DC, September 20, 2022.

9. Crystal McCoy, "Unjust Transition," *Earth Island Journal* (Winter 2022).

10. Environmental Protection Agency, Sources of Greenhouse Gas Emissions, last updated April 28, 2023, https://www.epa.gov/ghgemissions/sources-greenhouse-gas-emissions.

11. Joe Zimsen, "Invest in Iowa Clean Energy," *The Gazette*, May 3, 2022, www.thegazette.com/guest-columnists/invest-in-iowa-clean-energy/.

12. Personal communication with author, October 20, 2022.

13. Personal communication with author, October 21, 2022.

13. Solidarity for the Snowpack

1. Colorado Economic Overview, "Colorado—State Economic Profile," IBISWorld State Industry Reports, IBISWorld, 2022, www.ibisworld.com/united-states/economic-profiles/colorado/.

2. Colorado Office of Economic Development and International Trade, "Tourism," Choose Colorado, choosecolorado.com/key-industries/tourism/.

3. Colorado Office of Economic Development and International Trade, "Tourism."

4. Colorado Office of Economic Development and International Trade, "Colorado's Tourism Economy Saw Growth in 2021 According to Annual Research Reports," Colorado Office of Economic Development and International Trade, July 14, 2022, www.oedit.colorado.gov/press-release/colorados-tourism-economy-saw-growth-in-2021-according-to-annual-research-reports.

5. Carl Talsma, Katrina Bennett, and Velimir Vesselinov,

"Characterizing Drought Behavior in the Colorado River Basin Using Unsupervised Machine Learning," *Earth and Space Science* 9, no. 5 (2022).

6. Mark Cox, "Snowmaking Stretches Ski-Season. But Is It Sustainable?" *Red Magazine*, November 22, 2022, www.red.msudenver.edu/2022/snowmaking-stretches-ski-season-but-is-it-sustainable/.

7. Joseph T. O'Connor, "Big Ski Patrol Votes to Unionize," *Explore Big Sky*, May 5, 2021, www.explorebigsky.com/big-sky-ski-patrol-votes-to-unionize/38415.

8. Jonathan Romeo, "A Unified Front: Purgatory Ski Patrol Moves to Unionize for Better Pay, Benefits," *Durango Telegraph*, March 31, 2022, www.durangotelegraph.com/news/top-stories/a-unified-front/.

9. Brian Spieker, "PCPSPA Solidarity Fund," GoFundMe, December 25, 2021, www.gofundme.com/f/pcpspa-solidarity-fund.

10. Zac Podmore, "'We Meant Business': Park City Ski Patrollers Say Strike Authorization Vote Led to Better Contract with Vail Resorts," *Salt Lake Tribune*, January 20, 2022, www.sltrib.com/news/2022/01/20/we-meant-business-park/.

11. "Lift Mechanics and Electricians at Park City Mountain Resort, UT, Looking to Unionize," *Snowbrains*, October 11, 2022, www.snowbrains.com/lift-mechanics-and-electricians-at-park-city-mountain-resort-ut-looking-to-unionize/.

12. Peter Landsman, "Park City Lift Mechanics Vote to Unionize," *Liftblog*, November 22, 2022, www.iftblog.com/2022/11/22/park-city-lift-mechanics-vote-to-unionize/comment-page-1/.

13. "Park City Mountain Resort, UT, Maintenance Vote to Unionize," *Snowbrains*, November 23, 2023, www.snowbrains.com/park-city-mountain-resort-ut-maintenance-vote-to-unionize/.

14. More information is available at greenworkers.org and breachcollective.org.

15. Terry O'Sullivan, "DAPL Letter to Membership," *LiUNA!*, October 26, 2016, www.liunaactionnetwork.org/site/DocServer/2016-10-26_DAPL_Letter_to_Membership_Oct_2016_FINAL.pdf?docID=2721.

16. Ty Adams, Gary Arnold, Sara Blackhurst, and Rich Meisinger, "Opinion: Natural Gas Should Remain a Key Option in Colorado's Clean-Energy

Mix," *Colorado Sun*, January 25, 2022, www.coloradosun.com/2022/01/25/colorado-energy-natural-gas-electricity-opinion/.

17. University of Delaware, "New England Fishermen Losing Jobs Due to Climate Fluctuations: Direct Link in First-of-Its-Kind Labor Study," *Science Daily*, December 9, 2019, www.sciencedaily.com/releases/2019/12/191209161317.htm.

18. "Colorado River Drought Affecting Rafting Trips," Advantage Grand Canyon, www.advantagegrandcanyon.com/colorado-river-drought-affecting-rafting-trips/.

19. Lauren Mowry, "Will the Sport of Scuba Diving End By 2050?," *Forbes*, June 2, 2017.

14. The Fight for America's Workers Must Also Be a Fight for Environmental Justice

1. Paul Rogers and Paiching Wei, "Map: 1 of Every 8 Acres in California Has Burned in the Last 10 Years. Here's Where the Biggest Fires Spread—and Are Burning Now," *Mercury News*, September 19, 2021, www.mercurynews.com/2021/09/29/top-10-california-wildfires-megafires-map/.

2. Anna Phillips et al., "Heat Waves Are Far Deadlier Than We Think. How California Neglects This Ultimate Threat," *Los Angeles Times*, October 7, 2021.

3. R. Pollins, "California Climate Jobs Plan," California Climate Jobs, www.californiaclimatejobsplan.com/equitable-transition.

17. Listening to the Land, Listening to the Workers

1. Author interview with Anayeli Guzman, August 2022.

2. At North Bay Jobs with Justice, we believe that the Land not only sustains our work but is an active participant in our worker-led vision for climate justice. We choose to capitalize its spelling to recognize and honor its subjectivity.

3. Author interview with Maria Salinas, March 2022.

4. Author interview with Anabel Garcia, November 2022.

5. Author interview with Sandra De Leon, August 2022.

6. Quote from speech by Margarita Garcia at Simi Winery Picket, November 2021.

PUBLISHING IN THE PUBLIC INTEREST

Thank you for reading this book published by The New Press; we hope you enjoyed it. New Press books and authors play a crucial role in sparking conversations about the key political and social issues of our day.

We hope that you will stay in touch with us. Here are a few ways to keep up to date with our books, events, and the issues we cover:

- Sign up at www.thenewpress.com/subscribe to receive updates on New Press authors and issues and to be notified about local events
- www.facebook.com/newpressbooks
- www.twitter.com/thenewpress
- www.instagram.com/thenewpress

Please consider buying New Press books not only for yourself, but also for friends and family and to donate to schools, libraries, community centers, prison libraries, and other organizations involved with the issues our authors write about.

The New Press is a 501(c)(3) nonprofit organization; if you wish to support our work with a tax-deductible gift please visit www.thenewpress.com/donate or use the QR code below.